Information, Knowledge, Evolution and Self

by the same Author

Science:
The Dawkins Deficiency: Why Evolution is Not the Greatest Show on Earth
- published by Deep River Books, Sisters, OR, 2011

Fiction:
Finding the Shepherd - A Tale of Two Loves
- published by Westbow Press, Bloomington, IN, 2016

From the Back Pew
(series published by Peshat Books, *www.peshatbooks.com*)

Volume 1 - *If Not God What?* On Being an Intellectually Fulfilled Theist

Volume 2 - *Choosing to Know God*: Understanding God's Presence in the World

Volume 3 - *Bible Inerrancy: Fact or Fiction?* The Inerrancy of God's Word versus the Fallibility of Human Interpretation

Volume 4 - *Our Shepherd His Flock*: Following the Jewish Messiah on the Path Less Travelled

Volume 5 - *What New Covenant?* Rethinking the Implications of the First Coming of the Messiah

Volume 6 - *God's Only Law Book*: Overdue Maintenance of the Narrow Path

Volume 7 - *Defending God's Sabbath*: Obeying God's Commandment to Safeguard the Sabbath

Volume 8 - *From Sin to Salvation*: A Fresh Perspective on God's Plan for Mankind

Volume 9 - *The Four Pillars*: The Four Factors that Persuade me to Accept God and Dissuade me from Accepting Evolution

Volume 10 - *What God Commanded*: Love Him by Keeping His Commandments

Volume 11 – *A Biblical Discourse*, Volume 1 – For Those Prepared to Risk Their Orthodox Theology

Information, Knowledge, Evolution and Self

A Question of Origins

WAYNE TALBOT

Copyright © 2016 by Wayne Talbot.

Library of Congress Control Number: 2015920932
ISBN: Hardcover   978-1-5144-4422-1
             Softcover   978-1-5144-4421-4
             eBook   978-1-5144-4420-7

All rights reserved. No part of this book may be reproduced or transmitted in any form or by any means, electronic or mechanical, including photocopying, recording, or by any information storage and retrieval system, without permission in writing from the copyright owner.

Any people depicted in stock imagery provided by Thinkstock are models, and such images are being used for illustrative purposes only.
Certain stock imagery © Thinkstock.

Print information available on the last page.

Rev. date: 01/28/2016

To order additional copies of this book, contact:
Xlibris
1-800-455-039
www.Xlibris.com.au
Orders@Xlibris.com.au
732135

Contents

Dedication vii
Acknowledgements.. ix
Author's Note xi
Thesis... xv
Introduction xxi

Chapter 1: Does Knowledge Trump Evolution? 1
Chapter 2: Understanding Our Senses 9
Chapter 3: The Information Blind Spot 12
Chapter 4: Data, Information, and Communication 19
Chapter 5: Data Processing Principles 24
Chapter 6: Information Processing in the Mind 42
Chapter 7: The Irreducibility of Sight...................................... 49
Chapter 8: Knowing That You Can 64
Chapter 9: Volition and Free Will 75
Chapter 10: Distinguishing Mind from Brain 80
Chapter 11: Does the Brain Store Memories? 91
Chapter 12: Speaking, Writing & Reading 95
Chapter 13: Understanding "Self".. 100
Chapter 14: The Fallacy of Materialism 106
Chapter 15: The Reality of Emergence.................................... 109
Chapter 16: Conclusions 113

References 117
About the Author 123

Dedication

To my good friend Charley in California, a man to whom I look with admiration and wonderment at his unfailing courage and endurance, a man who better than anyone I have ever known, demonstrates the mysterious nature of the human spirit.

Keep up the good fight, Charley – you lead and inspire us all, for whenever I feel my spirits flagging, I think of you.

Acknowledgements

A foundational theme throughout this study is that all knowledge is built upon prior knowledge, *turtles all the way down* as it were. Though I have lacked the good fortune of contemporary collaborators, I am otherwise fortunate to have access to the knowledge of scientists and scholars going back to the beginning of written history, the very existence of which testifies to my conclusions.

Of more direct relevance is the research of luminaries such as Claude Shannon, Werner Gitt, William Dembski, and Walter ReMine, for their published works have been the trigger for my own. Counterintuitively, it has not so much been the content of their books, but what I have perceived as missing. I acknowledge that this is a backhanded compliment, but forgive me, for I mean no offence. In a similar vein, even evolutionists such as Richard Dawkins have contributed significantly to my thinking. In recognising the failure of logic in arguments supporting evolution, I have been prompted to seek the truth for myself.

I wish to thank all those who have trodden this path before me, for without their work, I would have had nothing upon which to build my own. Curiously, foundations both sound and unsound have proven to be useful.

Special thanks go to Michael Egnor, for his encouragement has meant more to me than I can express.

Author's Note

Though some readers may choose to see it as such, this study is not an attempted refutation of the over-arching narrative of evolution, as I understand it: *that all life on earth arose from a single common ancestor which itself arose from an inorganic form.* Similarly, some of my reference sources may suggest to the reader that I am arguing from the stance of intelligent design or even creationism, but such is not the case - there is no deistic, theological, or religious basis for any of my arguments. My goal in this study is to evaluate in the context of the evolution hypothesis, what is widely known from the sciences of knowledge and information processing, and the thoughts and experiences of philosophers down through the ages.

The term "evolution" is so elastic that any discussion in this area must be preceded by a definition in context. For example, when Professor Futuyma states that *"evolution is the single most pervasive theme in biology, the unifying theme of the entire science,"*[1] is he referring to *descent by modification* in its broadest sense, or more simply genetic inheritance generation by generation?

In his book, *Science on Trial – The Case For Evolution*, Professor Douglas Futuyma stated that Darwin drew *"his evidence from comparative anatomy, embryology, behaviour, geographic variation, the geographic distribution of species, the study of rudimentary organs, atavistic variations (throwbacks), and the geological record to show how all of biology provides testimony that species have descended with modification from common ancestors."*[2] In recent years, research into genetics and related fields has brought the basis of Darwin's conclusions into question, with some scientists now asking whether the essential mechanisms of evolution have been correctly identified. Published works by Jerry Fodor & Massimo Piatelli-Palmarini[3], Suzan Masur[4], Stephen Meyer[5], John Sanford[6], James Shapiro[7], Lee Spetner[8], and Robert Wesson[9] testify to the discussion. Putting that aside, but keeping in mind that these scientific fields of research were totally unknown in Darwin's days, there is to my mind a far more important field of research about which I have been unable to find any published works. That is not to suggest that such do not exist - simply that I have been unable to unearth them.

Of particular interest is a symposium held at Cornell University is the Spring of 2011 with the proceedings published in 2013: *Biological Information - New Perspectives*[10]. I have not studied this particular work, it being highly technical and initially very expensive, but I have reviewed *The Synopsis and Limited Commentary*[11] by Dr. Sanford to determine the degree of overlap with the material being presented here in my own study. There is considerable overlap; for example, Dr. Oller's paper on Pragmatic Information refers to "simple mathematical fact that the number of possible strings at any given level in any natural language, or any language-like biological signaling system, grows exponentially as we progress up the hierarchy of information layers" - this supports my arguments regarding data ancestors and that data only becomes cognitive information when processed through a referential framework of concepts providing context. Dr. Donald Johnson, with PhDs in both Computer Science and Biology, validates my comparison with digital information processing and storage when he "demonstrates that the information networks found within living cells are remarkably similar to computer networks". Dr. William Dembski, "widely known for his work in information theory and information search strategies", notes that "a search program cannot be designed to do any better than a random search, unless the designer has vital information on which to base the search". In a later chapter, I will discuss the concept of a "search space" which is framed by the search arguments, and how in cognitive as opposed to biological information processing, the wide gap between the conceptual and the physical cannot be bridged without an intelligence capable of cognitive abstraction. There is much more, much much more, such as the limits of self-organisation, but we will come to these in their proper place.

The take-away, if I can use that term, is that most of the principles that I discuss in this study have been validated by recognised scientific experts in relation to *biological information*, and thus there should be no rejection of them in relation to *cognitive information*. The important issue though, is that there is a substantive difference in the application of these two types of information, and it is this difference that I wish to illuminate in this book.

I have studied works by William Dembski[12], Daniel Dennett[13], John Eccles[14], Werner Gitt[15] [16], Martin Heidegger[17], Thomas Nagel[18], Denis Noble[19], Karl Popper[20], Walter ReMine[21], Gilbert Ryle[22], and numerous essays by lesser luminaries, but none treat the subject of information and knowledge in quite the way that I do here. I will leave it to the reader to adjudge whether I have added any new thoughts.

There are many variations, even in evolutionary terms, in the understanding of how life came to exist in its present forms. The materialmonist asserts that the evolutionary processes were entirely undirected, whilst others like the former head of the US Human Genome Project, Francis Collins, perceive Divine guidance and has offered a synthesis of science and theology[23]. Dr Denis Alexander, Director of the Faraday Institute for Science and Religion has written similarly[24].

This is not an arena that I choose to enter in this study. Though I am not a scientist in the accepted sense, I will nevertheless attempt to tread the path of the scientist in pursuit of the evidence that is readily available. In an earlier work[25], I sought to expose the errors in Richard Dawkins' *The Greatest Show on Earth*[26], but in this book, I seek to expose an area of study that to my knowledge, has not been pursued with any vigour: that concerning the *origins* of cognitive, as distinct from biological, information and knowledge, and the implications for evolution theory. There continues to be a great deal of work in the neurosciences, and energetic debate over mind (conceptual) versus matter (brain) as we shall see, but such is primarily in the domain of what is happening now rather than how it came to be.

Though the evidence and reasoning in this book have contributed to my personal conclusions regarding the origin of life, I would not presume that it should so conclude for others, for each of us should carefully weigh the evidence for ourselves.

Finally, this is not a science book in the accepted sense, for I am not a scientist. The target readership is people like myself, well educated in a number of fields, enthusiastic amateurs if you like, but willing and able to see through the fog of technical language and unsupported assertions to discern the truth for themselves. Of course, I would welcome readership amongst the scientific community, but such people should understand that some of the rigorous norms of scientific publications are absent.

Wayne Talbot
Kelso NSW Australia
2015

Thesis

Based on the evidence that I have been able to research, I contend that the undirected, organic mechanisms of evolution so far offered by the proponents of its over-arching narrative, cannot account for the *cognitive abilities* of humans. To constrain the context of the discussion I would offer two working definitions:

> *Cognition*: the set of all mental abilities and processes related to knowledge. These include the interpretation of sensory inputs, commitment to memory, organisation of memory, recall from memory, determination of relevance, rejection of "noise", correlation of abstractions, comprehension, invention, judgment, evaluation, logical reasoning, abstract reasoning, computation, problem solving, decision making, development and use of language, etc.
>
> *Evolution*: The theory of how complex organisms arose on Planet Earth. I use the definition as given by Professor G.A. Kerkut: *"the theory that all the living forms in the world have arisen from a single source which itself came from an inorganic form"*[27]; I will refer to it as the "General Theory of Evolution", or the over-arching narrative of evolution.

I acknowledge that this particular definition of evolution is not one used by evolutionary biologists and other specialised disciplines, but I consider it disingenuous, perhaps even intellectually dishonest, to speak of origins yet start somewhere in the middle. This study concerns the *origins* of information, knowledge, and the cognitive applications by the human mind. As I will attempt to demonstrate, the evolution narrative for all its twists and turns, inventions of hypothetical processes, and elasticity in definition, fails entirely in its attempts at explanation.

Abstraction

The word "abstraction" can be used in two senses, and it is well that I clarify my usage. To *abstract* can mean to *separate*, *extract*, or *distillate*, but that is not what I intend here. *Abstraction* as used in this study refers to the intangibles that result from cognitive processes and the mental conceptions so formed. Works of art, when experienced, result in different conceptions in different people, as do music, poetry, literature, scenery, and other physical entities. As is often expressed, "beauty is in the eye of the beholder": affirming that the perception of beauty is subjective. A point that I would make from the outset is that cognitive processes in general are of that nature - subjective. A primary issue in this study is to what do we ascribe the link between the physical (organic) and the conceptual (non-physical): can that be entirely organic or is there some as yet unrecognised immaterial process involved?

A related question, and one which perhaps lies at the very heart of the problem, is whether the purely organic, which is objective, can give rise to the conceptual, which is subjective?

Opening Thoughts

To understand this subject in relation to the overarching narrative of evolution, one must first understand a simple scientific fact, one that is generally hidden by imprecise language:

> Material storage devices, irrespective of whether they be synthetic or organic, do NOT and CANNOT store *information*.

The same is to be said for communications of any form, through any media. I realise that this may be difficult to accept, so let me explain. Firstly, the word "information" is the noun equivalent of the verb "to inform", and thus if a communication does not *inform* you, it is not *information* to you. It may be information to somebody else, but that is irrelevant in the context of an individual's cognitive processing and knowledge. In the context of evolution, we must always consider the individual organism, for that is where the mechanisms of evolution are said to occur before impacting a wider group.

Examining what are loosely termed, *communications*, we should understand that the letters on this page are symbols devised for a specific coding system. To decode them, the reader must have pre-loaded data relating to language, alphabet, vocabulary, grammar, syntax, punctuation, and knowledge of the subject matter. If the symbol set (alphabet) is familiar to you but the language is not, no useful communication occurs. If the symbol set is unfamiliar, such as the Hebrew or Cyrillic script, or even worse, Oriental pictograms, then again no communication can occur. I appreciate that this is so basic that it hardly needs restating, but we need to keep these basics in mind when examining how the acquisition of knowledge could have occurred through evolutionary processes for the very first time.

What may not be immediately obvious is that every one of these terms used above are "concepts", absent of understanding of which no useful communication can occur. A dictionary is useful for building one's vocabulary, but no dictionary can convey foundational concepts: these the reader must bring with them to the book. For example, my dictionary defines an *elephant* as a: "large pachyderm with proboscis and long ivory tusks". That hardly helps in understanding what an elephant is because there are at least four concepts which must be preloaded: pachyderm, proboscis, ivory, and tusks.

My point, on which we will elaborate as we proceed through the study is this: *data only becomes cognitive information when intelligently processed through preloaded concepts which provide context.*

Now let us have a closer look at "data". Material storage devices do not even store *data* - they store *symbols* – of a type appropriate to the method of storage. The storage and retrieval of such symbols must be based upon a coding system, itself several levels removed from the data that is intended to be represented. Examine any material storage device in terms of physical properties and the truth of this should be self-evident. Examine a printed page and in terms of physical properties, all you will find are ink-stains in what appear to be regulated patterns, but keep in mind that for you to comprehend that, you had to pre-know concepts such as ink and patterns. Examine a computer hard disk – with what? The only properties that you will discover are variations in magnetic regions which in themselves are meaningless. Examine a music CD and you will not find music, or pictures on a DVD. Examine the human brain and all you will discover are organic substances subject to the laws of physics and chemistry, but nothing that you can comprehend externally as information content. Activity can be sensed, but not the informational content or context of that activity.

As with reading the symbols on this page, the physical or chemical symbols need to be processed through a hierarchical framework of concepts and contexts before information can be presented in a human cognitive form.

Repeating the earlier observation by Dr. Oller regarding Pragmatic Information: it is a "simple mathematical fact that the number of possible strings at any given level in any natural language, or any language-like biological signaling system, grows exponentially as we progress up the hierarchy of information layers." Note the reference to "hierarchy of information **layers**" [emphasis mine]. We will come back to that concept a little later. Dr. Oller was speaking of genomic information, but the same holds true for cognitive information: given the exponential growth of possible paths as we proceed from raw data to cognitive information, the probability that this occurred by undirected evolution clearly approximates zero.

In this study, I will begin with data and information processing principles as I contend that irrespective of the medium in which such processes occur, the principles do still hold. The human brain is often compared to a computer, but I will later argue that it is much more than any computer that humans can devise, the principal reason why I believe that *artificial intelligence* will never achieve the reality espoused by science fiction writers and some researchers and prophets of science.

When scientists discuss information in the context of evolution, they invariably refer to the *biological* information in DNA, but such information is substantively different to the *cognitive* information seemingly processed in the mind or brain. Biological information is both stored and processed using the same physical medium - chemicals, but if indeed the brain does store information, the processing of it is several levels abstracted from the storage process. The genome is chemical in nature: the processing is based on the intrinsic properties of those chemicals when arranged in specific ways, and transacted in specific sequences. We might say that in a sense, the intrinsic properties of the genome allow it to "know" what it is on about. In cognitive processing however, there is no natural chemical or biological relationship between the properties of the storage medium (assumed to be the brain) and the abstract relationships that the stored symbols represent. Again in that sense, we can say that the stored information does not know what it means.

In cell development, the genetic instructions that specify the required relationships, and the processing sequence for the desired end results, are encoded in the chemical medium itself: absent of mutation or other errors, the genome can only do what it does. Cognitive processing is entirely different in that there is nothing inherent in the chemicals that can give rise to imagination, creativity, volition, abstract thought, artistry, and the other wonderful capabilities of the human mind. Thus there must be another agency at work that is not present in our DNA, for unlike the genome, *the mind can do much more than it is.*

A Simple Test

To set the scene for what is to follow, try this simple test. Take a clean sheet of paper (or electronic equivalent) and on it write one single, isolated fact or quantum of information that requires no other facts or information to comprehend it. It is my hope that after whatever period of contemplation, you will have concluded that no such single fact exists. For example, you might write the letter "a" but in doing so, you have subconsciously understood this to be a letter (concept) of the alphabet (concept), the latter being essential to providing context. You might even have realised that in order to be able to carry out the task, you had to pre-know certain other facts such as what is meant by writing, paper, pencil, and so forth. Just as importantly, *you had to know that you could*. This latter truth is significant in the context of evolution as I will attempt to demonstrate in a later chapter.

By Way of Explanation

The title of this study is "*Information, Knowledge, Evolution and Self*". It may seem an odd combination of concepts, particularly the inclusion of "self", but I intend to show that *self* relies entirely on *knowledge of self* and for that one needs *information*. All of these are abstractions of reality but must in fact be based on some reality, else we are little more than illusions. Echoing the words of Shakespeare from *The Merchant of Venice*, Act III Scene I: "If you prick us, do we not bleed? If you tickle us, do we not laugh? If you poison us, do we not die?" I am firmly convicted of my physical reality - I too bleed and laugh, often at the same time, but so far as I am aware I have yet to die. Similarly, I cannot but believe in an immaterial reality that allows me to recall such words at will, and apply the concepts contained therein to a separate field of endeavour such as these writings.

This raises the question: what is the linkage between physical realities and conceptual realities? Those committed to philosophical materialism will insist that the immaterial cannot influence the material, or in any way cause a material reaction, but their philosophy is contrary to our everyday experience. The intriguing issue here is why material monists would even proffer such an assertion when they deny the very existence of the immaterial. We will investigate these issues as we proceed and hopefully, by the time we get to the chapter(s) on *Self*, you will be at least part way to agreeing with me.

Introduction

There was a time in history when the term "science" covered a much broader spectrum of human intellectual endeavour, and many of the most notable scientific discoveries were by *polymaths* - people who had expertise in numerous subject areas allowing them to draw on complex bodies of knowledge including philosophy and yes, even theology. In more recent years, perhaps tracing its origins back to the *Enlightenment Era*, the term "science" is more commonly limited to the physical sciences as understood by the material-monists.

The fact that the General Theory of Evolution provides the best *materialistic* explanation of origins is not in itself reason to accept it unquestioningly, unless of course, one's personal philosophy compels them to accept only materialistic explanations; all others should keep an open mind. As Robert Wesson, evolution proponent, political scientist, and Senior Research Fellow at the Hoover Institute put it:

> "The contention that reality consists of only material particles and their modes of interaction is not even a clear-cut theory. It implies a narrow definition of reality, making the thesis true by definition: if only material substance is real, then material substance contains the whole of reality. But are the laws of nature not real? Are mathematical theorems real? Are patterns real? Are thought and consciousness? It is paradoxical to deny their essentiality, for science could not exist without them."[28]

Richard Lewontin, described as an evolutionary biologist, geneticist, and social commentator, had this to say:

> "We take the side of science in spite of the patent absurdity of some of its constructs, in spite of its failure to fulfil many of its extravagant promises of health and life, in spite of the tolerance of the scientific community for unsubstantiated just-so stories, because we have a prior commitment, a commitment to materialism. It is not that the methods and institutions of science somehow compel us to accept a material explanation of the phenomenal world, but, on the contrary, that we are forced by our *a priori* adherence to material causes to create an apparatus of investigation and a set of concepts that produce material explanations, no matter how counter-intuitive, no matter how mystifying to the uninitiated. Moreover, that materialism is an absolute, for we cannot allow a Divine Foot in the door."[29]

It is clear from these and similar quotes that many scientists suspect that there may well be more to our existence than can be described in purely materialistic terms. Not being equipped by inclination or training however, they prefer to not venture outside of safe harbour. Those committed to philosophical materialism claim to be pursuing truth, but it is as if they are faced with two doors both labelled "TRUTH", with one having a codicil in fine print that they are unwilling to read out aloud. They nevertheless choose that particular door.

Inference to Best Explanation

I will be so bold as to contend that there is not a single, substantive, scientific fact, nor substantive scientific evidence, for the General Theory of Evolution. Let me explain. Substantive evidence is that which supports just one proposition, and one proposition only. If the evidence can be used to explain more than one proposition, then it can only be circumstantial, and this is what we discover when we objectively evaluate the scientific evidence used to support the over-arching narrative of evolution. Whether from *microbes to man*, or inorganic *molecules to microbes*, substantive evidence has yet to be uncovered.

There is a wealth of <u>circumstantial</u> evidence evaluated within the accepted paradigm, and much hypothesising and reasoning leading to claimed *inference to best explanation*, but all of this is to the arbitrary exclusion of alternative explanations. There *is* a great deal of substantive scientific evidence for some evolutionary processes such genetic mutation, genetic drift, speciation, changes in allele frequencies, descent with modification, and even natural selection, but none of these can be said to offer a satisfactory explanation for the *origin* of the genome. Similarly, scientists have coined new terms for phenomena that are believed to have occurred, but for which the evidence is circumstantial at best, for such "evidence" resides only in the mind of the researcher trying to explain anomalies in earlier hypotheses. Terms such as *convergent evolution* and *emergent properties of the brain* are examples.

It should be understood that an inference to best explanation is always subjective, not objective, for any conclusion is subject to a person's worldview, personal philosophy, education, training, and the prevailing paradigm within which such enquiry is conducted. If a particular possibility is arbitrarily ignored, then our understanding of "best" must be similarly limited.

It is not my intention to disparage individual scientists or scientific endeavour in general, particularly as my entire life has been substantially enhanced by the advances in medicine and various technologies. My working life in aviation and later, Information Technology, was based on new technologies, but some things need to be said. Society has largely succumbed to two related phenomena: the *Cult of Personality* and the *Cult of Authority*. In the first, the opinions of celebrities are accepted without reservation simply as a function of their celebrity status, irrespective of whether their opinions are informed or not; in the second, the pronouncements of authoritative figures in various domains are accepted without reservation, even when just a smidgeon of discernment would caution otherwise. The problem is compounded when pronouncements are made by someone in both domains. It should be understood that many so-called scientific statements have no underlying science. For example, we have "experts" in alien life forms despite there being no evidence of alien life forms - how can there be experts?

We also encounter numerous self-refuting arguments by scientists, such as we will discuss in a later chapter on volition and free will. A recent example was the claim, still believed by some apparently, that our universe could create itself out of nothing. When you read past the headline, you encounter conditional statements such as "in the presence of" which is clearly the antithesis of "nothing". It would seem that such nonsense is acceptable when it comes from a person in scientific authority, but I would contend that we should attempt to keep scientific pronouncements in perspective. As evolutionary biologist, Professor Douglas Futuyma admitted:

> "In fact, scientists are just as human as anyone else. They believe that one or another hypothesis is most likely to be true, and they engage in sometimes bitter battles to defend their ideas. Scientists' beliefs are also shaped by their political, social, and religious environment."[30]

Concluding a discussion on how people can be misled, he went on to say:

> "Thus the common image of scientists as abstracted, unbiased, detached intellects has no foundation in reality. Scientists are often highly opinionated, even in the face of contrary evidence; and they are often not particularly intelligent either. The spectrum of scientists, as of any other group of people, runs from the brilliant to the fairly stupid."[31]

Lest anyone should feel offended by this gentle tirade, the best advice I can offer is that if the shoe fits, then wear it, but if not, these criticisms are not aimed at you. I simply wanted to offer a warning that the status or qualifications of a speaker are not necessarily a reliable guide to truth. Let me give a simple example on a subject much in the news: the search for extra-terrestrial intelligence and the planets which may host such beings. There have been pronouncements by astronomers of *Earth-like planets*, but upon analysis of the characteristics, they are anything but. For example, a planet without spin, with one side continually exposed to the heat of a sun, and the other exposed to the cold of deep space, is not earth like, and if it has an atmosphere, then the pressure differential as a result of the temperature differential would cause continuous gale force winds. Hardly Earth-like in the sense of the surface being conducive to life and habitation.

Background Research & Initial Thoughts

Let me admit that I have only *sampled* the available literature rather than surveyed it comprehensively. I have studied some related works as noted earlier, and have dabbled in the works of earlier philosophers such as Descartes. In the Information field, I am well acquainted with the works of Claude Shannon, Werner Gitt, Walter ReMine, and many others. In the context of my study here, I would contend that to a large extent, these very capable experts have nevertheless largely focused in specific areas leaving aside some fundamentals. That is undoubtedly presumptuous of me so let me explain.

Some years back, I was fortunate to have studied under a Cognitive Psychologist, Dr David Taylor. Dr Taylor had developed a business engineering methodology that he termed, *Convergent Engineering*[32], intrinsic to which was an analysis method called *Responsibility Driven Design*. In my thirty-five years in the industry, I found no approach to be as productive, and have continued to use the principles I learned from Dr Taylor in many other cognitive tasks. I notice that Dr Taylor's ground-breaking work was later extended by Richard Hubert[33]. I mention this both in acknowledgement of Dr Taylor's contribution to the development of cognitive skills, and to identify my analysis methodology. As an aside, some evolutionists claim that we have no free will and thus Dr Taylor could have had no personal responsibility for his work - it was all just the result of undirected complex chemical reactions in his brain. Knowing Dr Taylor as I do, I am confident that he would be quite offended by that suggestion, but we will come back to that later.

In his book, *The Concept of Mind*[34], Gilbert Ryle stressed the distinction between knowing *that* and knowing *how*. Those researching the cognitive sciences are aware of our cognitive abilities (knowing *that*) but are still mystified as to the mechanisms (knowing *how*). To know how, we need to deconstruct the process much as one would reverse-engineer a complex product to understand how to construct new ones, the goal of AI (Artificial Intelligence) proponents. We cannot do that if our presuppositions arbitrarily exclude avenues of research, particularly those related to origins. Cosmologists generally accept the Big Bang model for the origin of our universe, but due to inconsistencies in the science, some scientists are researching other origin solutions. We have the same issue with evolution of life: *abiogenesis* is accepted as fact even though there is (as yet?) no plausible scientific explanation, and thus some scientists propose alternate origins such as *panspermia*. Of course, this is hardly a solution as it simply pushes the problem onto another planet. If the posited singularity of the Big Bang is not true, and abiogenesis is not true, then how can our ideas on how the universe started and how life on Earth started be true? Scientific endeavour generally starts with what *is*, rather than what *was* (origins), a natural process and arguably the most productive, but if we entirely ignore the science of origins we cannot be sure that current research efforts are not heading down a blind alley. Starting an investigation in the middle, whilst perhaps necessary initially, leaves one open to unsupported assumptions; through convenience and repetition, such eventually becomes accepted truth.

The point that I am driving toward is that before we can investigate the essence of cognition, we need to differentiate the *conceptual* from the *physical*. In the works of Daniel Dennett and others that I have studied, I have found no attempt to do so. Authors often attempt to explain what they believe to be purely organic processes in anthropomorphic terms, as if teleology was a prerequisite, whilst at the same time denying teleology in evolution. This raises the question, as one article puts it:

"If it is not possible to speak of evolution's course without resort to the language of agency, is that a defect in human intelligence or an apprehension of fact?"[35]

There is an irony here which should not be missed. In his interesting book, *Consciousness Explained*[36], Daniel Dennett proposes a set of conceptual solutions without explaining how such concepts could be implemented at the physical, or what we might term the *bits & bytes* level. His computer analogies fall far short of what would be required for a plausible hypothesis. But you see, this is precisely the problem of cognitive processing: how can concepts be stored and retrieved at the physical level by a physical storage medium which has no intrinsic ability to comprehend the conceptual? Authors continue to offer conceptual solutions to physical problems, not realising that their struggle evidences the depth of the mystery. In the following chapters, I will explain the functional requirements of cognitive processing using a detailed analysis of digital information processing. I wish to thank Dr David Taylor for his insights into *Responsibility Driven Design* which have enabled me to see these issues clearly.

One further point before we begin. Quoting Daniel Dennett on the subject of dualism, he stated:

> "THE CHALLENGE: In the preceding section, I noted that if dualism is the best we can do, then we can't understand human consciousness. Some people are convinced that we can't in any case. Such defeatism, today, in the midst of a cornucopia of scientific advances ready to be exploited, strikes me as ludicrous, even pathetic."[37]

I read somewhere that *a conclusion is where you stop thinking*. I would offer that this applies to the material-monist who has concluded that the material is all there is, and that there is no point looking beyond the material. Thus we have the irony, dare I say hypocrisy, of somebody accusing others of defeatism whilst he himself has already embraced it, in this context at least.

CHAPTER 1

Does Knowledge Trump Evolution?

"Knowledge can be communicated, but not wisdom"
~ Hermann Hesse ~

There is a branch of philosophy which has been studied for at least 2500 years, predating Socrates, Plato, Aristotle and their ancient Greek contemporaries. More modern contributors include Bertrand Russell, Gilbert Ryle, Edmund Gettier, and Richard Kirkham. Their understanding of this subject should have been a red flag to Charles Darwin and all those who have since followed his ideas on evolution, but apparently the penny is yet to drop.

That subject is *epistemology*: a branch of philosophy in part described as concerning the nature and scope of knowledge, what it is and how it can be acquired. Closely associated is the philosophy of foundationalism which concerns philosophical theories of knowledge resting upon justified belief or some secure foundation of certainty. Intellectuals like Aristotle and Descartes wrestled with these ideas and realised that to avoid infinite regress in questioning our beliefs, one has to establish a starting point, and thus we have the term: foundational epistemology. In brief, this states that everything that we claim we know is based on something earlier known or believed, irrespective of whether it is factual or suppositional, true or false.

There are variations on this idea of foundationalism: classical, modest, internalism, externalism, and so forth, but I have yet to find any proponent for the idea that knowledge can come from nowhere at all. Thus there is a natural corollary to these ideas which has particular application to the theory of evolution: all knowledge must start somewhere and it cannot arise or create itself out of nothing.

As was mentioned earlier, Gilbert Ryle highlighted the distinction between knowing *that* and knowing *how*. He further argued that a failure to acknowledge this distinction leads to infinite regress: if all knowledge is built upon prior knowledge, where does knowledge start? This I believe is where we find evolutionists futilely burrowing into physics, chemistry, biology, and every other refuge of materialism. Borrowing from Ryle (but not in the sense of his argument): evolutionists believe that they know certain things inferred from circumstantial evidence, but do not know how such things came about. In the context of knowledge and the development of skills, we shall come back to the insights offered by Ryle.

Another very important concept explored by philosophers is that of *certainty*, which has led to other lines of enquiry such as the foundations of mathematics, inductive reasoning, probability interpretations, and Gödel's incompleteness theorems. We do not need to revisit the details here: we just need to acknowledge that whilst certainty is related to knowledge, it is fundamental to truth and reality. I am introducing a number of apparently unrelated concepts here but I beg your indulgence: I will draw the threads together as we proceed.

American philosopher, Edmund Gettier, contended that while justified belief in a true proposition is necessary for that proposition to be known, it is not sufficient. He also argued that there are situations in which one's belief may be justified and true, yet fail to count as knowledge. Here I am going to differentiate between true knowledge and false knowledge: not everything (information) we believe to be true is actually true and is thus termed false knowledge. I will persist with this differentiation of truth, information, and knowledge.

So far we have briefly reviewed some philosophical contentions, but now let us apply those to matters of science.

Science and Truth

Science, as we will use the term here, is concerned with the truth of our material existence and as we know from history, there have been and are proponents of both true and false knowledge. Scientific findings, when validated, can be used in technology and other practical applications, but when falsified are hopefully discarded. In between are numerous theories and hypotheses which have not been validated, and in some cases cannot be falsified. However, irrespective of our scientific views of what may be true or false, the material universe can only operate on what is true: if we observe some behaviour which is contrary to, for example, the laws of gravity, rather than pretend that it is not happening we should re-evaluate our understanding of gravity or some related phenomenon. This is what happened with the hypothetical planet, Vulcan, thought to be orbiting between Mercury and the Sun and causing the peculiarities in Mercury's orbit. Albert Einstein came up with new physics to explain the phenomenon without reference to a hypothetical planet, which in truth does not exist.

Something similar is happening with our understanding of gravity. It has long been understood as an attractive force but its nature and mechanism have yet to be determined. An alternate view of gravity has been proposed by Dr Russell Humphreys[38]: that space is truly a fabric, and every object having mass distorts the fabric such that they tend to "roll" toward one another. This is a metaphor for the reality, but it does help to visualise the phenomena: Dr Humphreys has proposed mathematical equations to explain the idea. My point here is simply that where our current scientific understanding fails to adequately explain phenomena such as gravity or the singularity that gave rise to the Big Bang, it is encumbent upon scientists to seek new understandings.

Technology aside, it is axiomatic that science and truth are always coincident in nature, irrespective of whether scientists have discovered such truth. Thus we can say that certain truths, and the information about those truths, pre-exist our own existence and that of any life forms which evolutionists claim gave rise to our own. Evolutionary biologists assert that abiogenesis is outside their field of enquiry and allowing that, if we wind back the clock to any point that evolutionists may choose, there is a great deal of truth that existed prior to it being acquired in the form of information and knowledge. The question then is: how can organic evolution account for our acquisition of knowledge?

As an aside, I am saddened when evolutionists such as Richard Dawkins assert that religion and theology threaten our knowledge of existence. Scientists claim to be searching for truth, but their claim is false if they refuse to look in places where unpalatable truths may be lurking. When scientists assert without looking that there is no evidence of the non-material, they argue from ignorance, for how could they know? Rather than being a source of knowledge and progress, such scientists are an impediment to knowledge and one wonders whether they feel threatened by what they choose to ignore.

Information in Humans

The study of genetics has generated an enormous interest in the information contained in the genome, but I would offer a refinement to the normal usage of the word "information". The coding of DNA is represented by the four letters: A, C, G, and T. Early studies thought that meaning could be derived directly from the sequence of these "letters" and whilst that is true to some extent, it is considerably more complex than that. Because our genes are both pleiotropic and polygenic, there is not a one-to-one correlation between a coded string and a protein. Thus those four letters at best represent data, not information, as I will explain in a moment.

I would contend that the information conundrum in the genome is far more easily solved than another occurrence of information in humans: that contained in our minds (or even brains for those who believe that the two are equivalent). Organic information in the genome is comparatively simple to understand because it behaves subject to the laws of physics and chemistry: even when the genome misbehaves, it still does so according to those very same laws. Because science and truth are coincident in nature, certainty exists in material behaviour even though our lack of knowledge and understanding may suggest otherwise. Scientists will often speak of random behaviour, but I suspect that the physical world never behaves randomly: it just appears that way due to our lack of understanding of the underlying causes.

The behaviour of our brains falls into two categories: *autonomous* and *volitional*. The autonomous functioning is pre-programmed at some stage of embryonic development although how that arose through evolution is unknown. The volitional functioning is entirely different and is even less understood, if at all. Curiously some evolutionists have claimed that we do not have free will and thus the very concept of volition is invalid. We will come back to that in a later chapter.

The Science of Information

Today we have a considerable advantage over philosophers of past generations: the advent of the computer has enabled us to manipulate vast quantities of data in ways never before imagined. Recent trends have seen the emphasis shift from the metaphysics of information to the technology, as seen in the works of Claude Shannon, Werner Gitt, and Walter ReMine: necessary but not sufficient in my view. Let me apologise in advance if I am in any way misrepresenting the very valuable achievements of these scientists, but my issue is the apparent failure to differentiate between data and information, and the consequent absence of an explanation as to how information originates and/or is derived. This has direct application to the feasibility of organic evolution of information, knowledge, and truth.

Claude Shannon is often described as "the father of information theory", but with no intent to belittle his ground-breaking achievements, more correctly we should say that he fathered the process of digital information *processing*. Shannon worked on the technical aspects of information processing with little regard for the truth or validity of the data being processed, and thus has contributed little to our understanding of the intrinsic nature of information. Werner Gitt has added to our understanding with his five-layer model[39] of Apobetics,

Pragmatics, Semantics, Syntax, and Statistics, but in some respects has overlooked a vital issue. In later works he has expanded his explanations to include the concept of *Universal Information*, but in constraining his thoughts to his earlier model, he has ignored the simple fact that some information flows are unintentional and do not conform to his model. In doing so, he has, like Shannon, contributed little to our understanding of the origins of information and its relationship to evolution. Walter ReMine has taken a novel approach with his concept of the *biotic message*[40], but this relates primarily to the genome and says little about cognitive information. Like evolutionary biologists, these scientists have started in the middle based on certain presuppositions. It is my intention to seek further back in time, looking for what might have been - not what is.

An "information" flow that I have not encountered in the literature is one that occurs in nature, and which few seem to recognise until pointed out to them. Its relevance to evolution is that for evolution to be true, it had to have been the *very first source* of information and knowledge for the evolving organisms. I speak here of data acquired by the senses, particularly sight and hearing. As I will discuss in a later chapter, a beam of light contains data pertaining to its source, and when reflected from an object, contains data pertaining to that object. That is the basis of our sense of sight. The issue is that the data is coded using a method appropriate to the medium of communication, and thus the communication contains neither data nor information, but merely physical symbols. For it to become data, it needs to be decoded, and to become information, the data needs to be processed against a hierarchical structure of concepts which provide context. If there is nothing there in the first place, how does the process get started?

Just as it took scientists considerable research to understand the data in the light beam itself related to its source in terms of the spectrum, we ought to consider the prerequisites for understanding the data related to the object from which the light is reflected.

Proposed Axiom

It is my contention that data only becomes cognitive information when processed within a referential framework of concepts which provide context. This should be obvious but sometimes we need to remind ourselves.

Take any single letter, word, number, or symbol, and in isolation it could mean anything - or nothing. How many readers recognise this symbol "ט" as the letter *tet* of the Hebrew alphabet? At an expanded level, consider finding a piece of paper inscribed with what appear to be Chinese or Japanese pictograms. Without a context, it may or may not be meaningful to you, but if the same characters were written on clean piece of 6x8 white cardboard, nestled in a plastic frame on the table of a Chinese restaurant, you would assume it to be a menu. Knowing that a menu is a list of foods, one then needs to understand both the categories of foods and the culinary variations by culture. Depending on the language in which the menu is written, you may still need to rely on the venue to provide the context. Nothing can be known until first the concept is explained, and then the context is applied to derive understanding. As the philosophers of old have long understood, knowledge is built upon knowledge, or as one lady remarked to Bertrand Russell in another context: "turtles all the way down". Russell responded by asking what was holding up the giant turtle, and we should ask the same question related to knowledge: what is the source of foundational knowledge?

In the early days of commercial computing, analysts and file designers unearthed a truth whose wider application is generally not recognised by modern practitioners. We will discuss these in a later chapter but here let me say that these fundamentals hold true for all forms of data or information processing, whether synthetic or organic.

What I wish to emphasise at this stage is simply this: all forms of communication and data storage are at the physical symbol level only. To be understood as data, the symbols need to be organised in a regulated manner known as *coding*. To derive an understanding, the recipient needs to have foreknowledge of the coding system. When decoded, the data still needs to be processed through a preloaded, hierarchical, referential framework of concepts before information can be derived. It is my contention that the agency that does the formulation of the coding system, the organisation of the symbols, and the search algorithms used for retrieval, must be external to the physical symbols themselves.

CHAPTER 2

Understanding Our Senses

From time to time, we encounter a new smell, a new taste, or an unfamiliar texture under our fingers. We are aware that these sensations are new to us, we can differentiate them from other sensations, but we cannot identify them. To do that, we need the services of an external agent that informs us via our other two senses, sight and hearing. Absent of that external agent, we remain in the dark, metaphorically speaking.

Now if we roll back the evolutionary clock to a time when the first sense supposedly arose, we have a problem. Whatever sense it was, all that it could communicate to the brain was an electro-chemical signal. Of course, the brain never having had one of those before, it had no idea of what to do with it, and no reason to store it. But let us assume that somehow the chemical properties of the embryonic brain were such that the sensation left an imprint – just where is anyone's guess. As further sensations arrived, the brain could perhaps differentiate them by their physical properties, but the storage location of the imprint would most likely be random. One could even ask why subsequent sensations did not overwrite previous sensations arriving via the same nerve?

Not only did the brain not understand what external reality was represented by the sensation, it did not know the source of the sensation in terms of the specific sense. For the brain to know that it was a smell, taste, or touch, it would have to know that such senses existed. The issue is that "sense" is a concept that has no physical reality until instantiated in a specific instance.

Let me explain.

In mathematics, there is no such entity as "a one" or "a two" – you can have "one of" but not "a one" because "one" is a mathematical concept, not a real entity. Similarly, the notion of *name*: name has to have an instance before it becomes "a name". There is no such physical entity as an unnamed name, just as there is no music without notes, and no songs with words. Number, name, music, and song are concepts which have no physical reality until instantiated. The reverse is logically equivalent. Smell, taste, and touch are physical realities, but the concept does not exist cognitively until the intervention of an external, intelligent agent to label the phenomenon. Smell, taste, and touch cannot self-identify as such: they simply exist as physical realities without self-awareness. Before any verbal or written communication can occur between humans, the concepts to be communicated have to be pre-agreed, otherwise we would all just be like politicians - talking past one another.

The other point to understand is that in those very early times in the evolutionary pathway, the brain could not know from where the sensations came. The brain might have been bombarded by cosmic radiation for all it knew, as all it was experiencing were unidentified sensations. The questions we must ask are these:

1. What are the distinct properties of the electro-chemical transmissions from the senses to the brain that differentiate them as coming from one sense as opposed to another?
2. If the circuitry of the nervous system is such that the source can be identified by the pathway, much as a computer can identify a port, what was the process that gave rise to the concepts of smell, taste, touch, sight, and hearing? Given that the brain could differentiate in some manner between the senses, it could still not identify them as to purpose or function, and in what way they represented the external reality.
3. Accepting that the embryonic brain could have differentiated between the senses as the source of the sensation, how could

it understand the variations in any one sense as representing specific characteristics of its external environment? It might "know" that variations existed, but it could not know what those variations represented. For example, the sight sensation could not know that variations represented colour, texture, distance, or size, because whilst these are physical realities, the cognitive concepts to describe them can only be derived by an intelligent agent. Similarly, "sweet" or "sour". Realistically, given that these evolutionary processes were supposedly the result of random mutations, it must have been pure luck that sight was not identified as sound, and sound, taste!

The brain's only sources of data, and thus information and knowledge, are the senses which the embryonic brain did not comprehend in terms of function, and in what way the sensations represented the organism's external reality, or even its own reality for that matter. Cognitive descriptions of self and the environment are conceptual in nature, whilst sensations are physical. The only bridge between the two, in either direction, is an intelligent agency. Cognitive intelligence, and cognitive processing, are at the conceptual level not the physical, even though neuroscientists have been able to identify locations in the brain where cognitive activity "appears" to be occurring. Before any observer could decode the physical activity, he/she would have to know the code structure, which itself would require the observer to perform cognitive functions based on prior knowledge not obtained from the events observed.

Quite obviously, if the path between the physical activity observed in the brain, and the cognitive concepts expressed, involve some form of coding structure, we would have to ask how such a coding structure arose in the first place without the intervention of an intelligent agency. In short, how can cognitive processing (intelligence) which is conceptual, arise from the purely physical, when the only bridge between the two is intelligence?

Enough on the senses for now as we turn our attention to the fundamental nature of information, as opposed to data and the symbolic representations in physical storage media.

CHAPTER 3

The Information Blind Spot

Science is focused on the technology of information as demonstrated by the work of Claude Shannon in particular, but that field is entirely separate from the study of cognitive information, and thus represents a significant blind spot in the science. As noted earlier, cognitive information is entirely different to genomic information is that it deals with concepts, whilst the latter deals only with physical entities. The brain, being organic like the genome, can only natively store physical entities, not concepts or abstractions. Concepts are derived from information in the form of knowledge, and as is posited in the science of epistemology, all knowledge is based on prior knowledge.

In an early draft of this paper, I coined the term *Conservation of Information*, but then encountered it in another context and thus had to abandon it. However, some aspects are common and I would to make a brief note of them here.

William Dembski commented: "Conservation of information is a term with a short history. Biologist Peter Medawar used it in the 1980s to refer to mathematical and computational systems that are limited to producing logical consequences from a given set of axioms or starting points, and thus can create no novel information (everything in the consequences is already implicit in the starting points)."[41] I believe the same to be true of cognitive information: everything that we derive conceptually was already implicit in our prior knowledge. We can create no novel information or knowledge ourselves - everything we know was already present as information before we were aware of it. In the context of evolution, and in passing, artificial intelligence in robots, where was this information before there were organisms capable of being aware of it?

Again quoting from Dembski's article:

> "Something unavoidably subjective and teleological seems involved in search. Search always involves a goal or objective, as well as criteria of success and failure (as judged by what or whom?) depending on whether and to what degree the objective has been met. Where does that objective, typically known as a target, come from other than from the minds of human inquirers? Are we, as pattern-seeking and pattern-inventing animals, simply imposing these targets/patterns on nature even though they have no independent, objective status?
>
> Mathematically speaking, search always occurs against a backdrop of possibilities (the *search space*), with the search being for a subset within this backdrop of possibilities (known as the *target*). Success and failure of search are then characterized in terms of a probability distribution over this backdrop of possibilities, the probability of success increasing to the degree that the probability of locating the target increases."

These observations can be applied to cognitive processes as well. People speak of "searching their memories" when asked "do you recall", and the analogy seems apt. But what does it mean in terms of cognitive processes: how do we search our memories accepting for the moment that there are such entities as "we"? Is that a volitional process or autonomous? When witnesses respond: *I can't remember, I do not recall*, or *I have no memory of that*, is their response volitional or autonomous? How then do we discern truth from falsehood (apart from the usual jokes about politicians and lawyers)? The article above states: "Search always involves a goal or objective, as well as criteria of success and failure" - by what criteria do we judge the success or failure of our mental search, particularly given that we sometimes recall incorrectly? Is the wrong answer solely a function of a fault or corruption in our neural networks, and if so, at what point in the cycle did the fault occur: during the storage process, whilst in storage, or during retrieval? Before we can begin to answer those questions, we need to have an understanding of the information processing cycle at a functional level.

The concept of *search space* has particular application to the cognitive processes of recognition and even creativity. A composer of music "searches" within a space of possibilities contingent upon culture and preference. Eastern music is often characterised by monophonic melodies, the melody being composed with reference to one note, most usually the tonic (first note) of the scale chosen. Western music tends toward polyphonic harmonies, where the melody is composed in reference to a chord or chord progression. Thus the *search space* of the composer is dependent upon the type of melody being composed, but let us not forget that the information in the search space must first be acquired.

Reading requires a search space firstly delineated by language which limits our search for vocabulary and grammar. Many words in most languages have a *semantic range*, i.e. a range of meanings, requiring context of the text to be evaluated in order to select the right meaning. The larger our vocabulary, the larger our search space and the greater our chance of success in understanding the text. However, vocabulary is also culturally conditioned: a text written in the 19th century may not be comprehensible to a modern reader, and most certainly, text language from Twitter or other social media would not be comprehensible to a 19th century reader. As the search space grows, the organisation of the data becomes more critical lest the search rapidly approaches the "needle in a haystack" conundrum. Actually it is worse – first you must know what a needle is.

Just as important as the *search space* is the *search algorithm*: how does the searcher know how to search the search space? In digital information processing, the storage process is intelligently devised, as is the search algorithm, without which any search is purely random with no opportunity to verify that what was found was the right answer.

A more fundamental example from the perspective of our sense of sight is given in the following image. Some people will "see" a rabbit, whilst others will not. The pertinent fact is that before anyone can perceive the rabbit, they must have prior experience of what a rabbit looks like. As we will investigate in a later chapter, this raises the question of how the very first organism to see a rabbit could recognise it as such, or in fact anything at all, without having a search space preloaded with comparative data. I will argue that in relation to our senses, the senses themselves are incapable of undirected acquisition of information and thus knowledge.

Irrespective of the physical implementation of information processing, there is a set of sub-processes which can be identified by their functional responsibilities. The following diagrams provide simplified overviews which we will discuss as the study proceeds. Please note that there is no intention of representing any formal architecture, and that the diagram notation is not based on any formal standard (quite deliberately so as not to be a point of contention). At this stage, I would like to draw your attention to the three processes numbered P1, P2, P3 in Figure 1. What may not be immediately obvious is that these processes, just like computer programs, must be stored somewhere and must be invoked in the correct sequence by a supervisory process. As an aside, early IBM operating systems were called *Supervisors*. In trying to explain consciousness, this concept of a supervisor has occupied the minds of philosophers (and scientists) down through the ages, leading to the proposition known as *dualism*: the position that mental phenomena are, in some respects, non-physical, or that the mind and brain are not identical. Material-monists are disdainful of this concept as they must be, for it is antithetical to their worldview, but the alternatives that I have encountered are far from convincing.

In the diagrams below and subsequent chapters, let me stress that I am not asserting that such processes must exist *per se*, but that the functions that they would perform need to be accomplished in some manner. Later chapters will elaborate on these functions and why they are necessary.

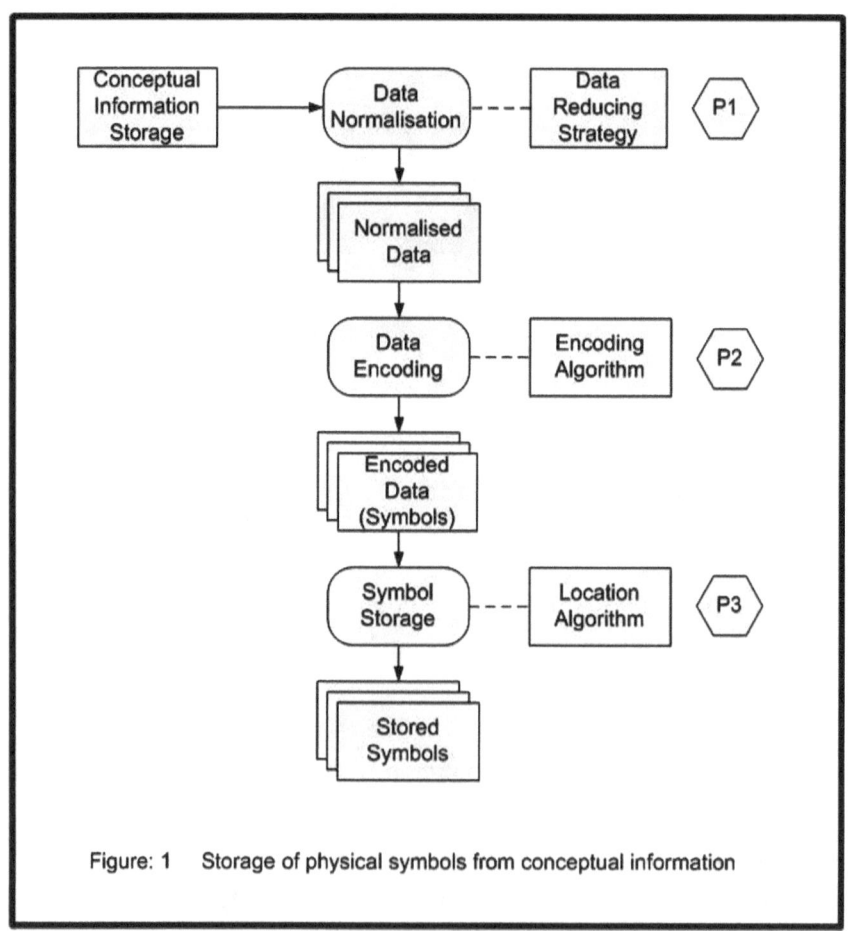

Figure: 1   Storage of physical symbols from conceptual information

Figure 1 provides a simple overview of the functions necessary to store conceptual information in a physical medium, irrespective of whether such is organic or synthetic. The first process (P1) is invoked to deconstruct the conceptual information into its component concepts, thereby removing the context: in computer language this is termed data normalisation. Process (P2) encodes the data appropriate to the method of storage followed by process (P3) which determines where the symbols are to be stored. The Location Algorithm is particularly important because it must "remember" where it put the symbols. Neural networks are of little value if the causal agency "forgot" where the symbols were stored and thus cannot locate that which it is seeking.

This brings us to Figure 2 which shows the functions necessary for the retrieval of conceptual information. As we saw in Figure 1, conceptual information consists of multiple component concepts, but not all are relevant to the search strategy. The components must be evaluated for relevance, and then encoded to match the storage format, again using P2. Process P3 which "knows" where things are stored is invoked to service the request and the requested symbols are returned. They are decoded using P5 which necessarily uses the reverse of the logic in P3. From there the retrieved data (concepts) are assembled contextually using P6 to form the requested information.

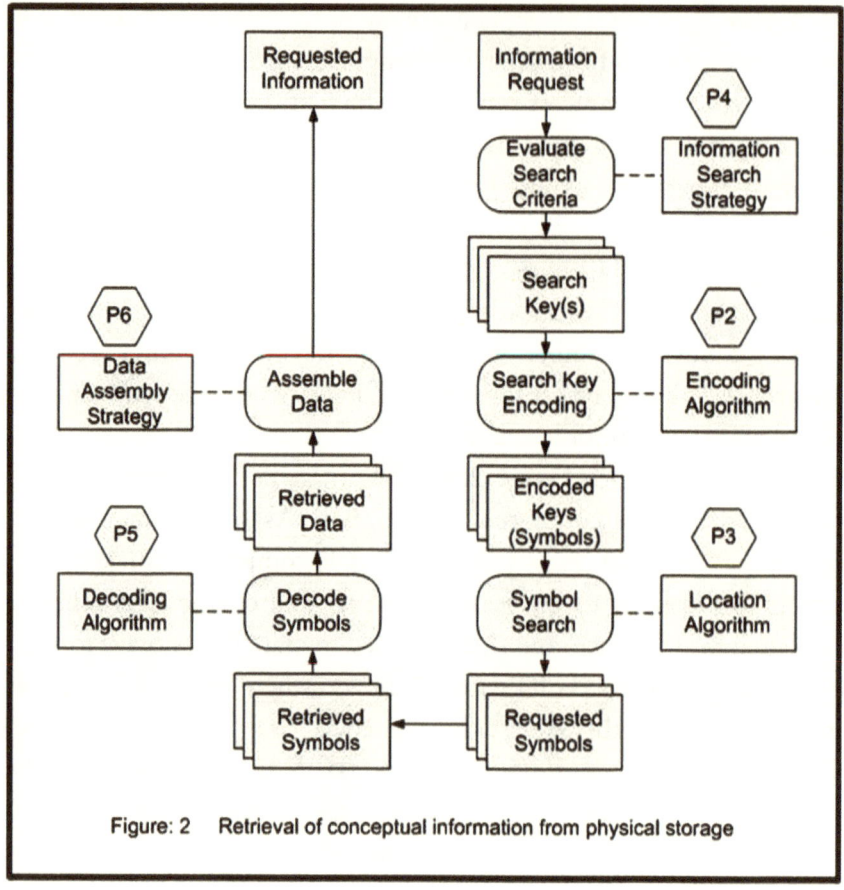

Figure: 2 Retrieval of conceptual information from physical storage

Our next step is to understand how this is accomplished in digital information processing systems, the point being to understand the functional requirements without suggesting what the biological architecture should be - just what it needs to achieve.

CHAPTER 4

Data, Information, and Communication

An insightful and concise definition of information is: *that which informs*, but in a sense it is inadequate because absent is a statement of who or what is being informed. Without a recipient, it is quite simply a broadcast rather than a communication, and thus there is no information flow as such. If your only language is English and you are handed a menu in Spanish, the menu does not contain *information to you* because you are unfamiliar with the coding system (language). If the menu is in Chinese or Russian, it is not information to you because you are unfamiliar with the symbol set (alphabet). There are clearly two separate layers of coding in the protocol: (1) the symbol set, and (2) the language. These fundamentals apply to any form of communication. Beyond that are vocabulary, grammar, punctuation and so forth but we need not go there at present.

Allow me to stress this point because it is essential to understand the differentiation between the technology of information and the cognitive processing of information, particularly when discussing information and knowledge in the context of cognition and evolution. Researching definitions of "noise", I find none that satisfy me in this context so I will coin a new term, *cognitive noise*, which I will hereinafter refer to as *humbug*, a term chosen for its obvious lack of formal scientific basis. When archaeologists first opened the tombs of the pharaohs, what was written on the walls was obviously information for the ancient Egyptians who wrote it, but *humbug* to the archaeologists who were unable to discern the meaning. So it is with all quanta of what is generically referred to as information: it is information to those who understand it, and *humbug* for everyone else. I would ask you to keep that distinction in mind for everything that follows.

Data transforming into Information

If the menu does not contain information (to you), then what does it contain? The Spanish menu may be said to contain data in that the alphabet is familiar, but the arrangement of the letters is not. The Chinese menu contains not even data but mere symbols, much as archaeologists experienced when first encountering Egyptian hieroglyphics or Sumerian script. All technical papers and journals that I have read on Information Technology, even those by eminent scientists and scholars, generally gloss over these important distinctions. To get from a symbol set to data, and from data to information, requires a cognitive observer irrespective of the method of transmission. But there is a prerequisite of even greater importance: the cognitive observer must have foreknowledge of the concepts and context of the communication.

For example: *"Friedrich's calculations showed that the particle involved in the exocytosis had to be orders of magnitude smaller than I had calculated from Margenau's equation"*[42]. I have no idea of what that means in isolation, but I do have some understanding from studying the complete book from which this excerpt was taken. I use this passage to illustrate a point: without foreknowledge of the concepts within a specified context, there can be no information exchange. Here is another interesting passage from the same book: *"For the first time, I recognised the fundamental significance of Akert's beautiful paracrystalline structure of the presynaptic vesticular grid of the synaptic boutons with its low probability for quantal emission of transmitter in exocytosis."*[43] Despite the explanatory diagrams and text, I lack the education (information and knowledge) to fully comprehend the significance of what is being said here. What I find quite incredible is how we humans can derive some understanding from texts when the vocabulary and abbreviations are beyond earlier education, but we nevertheless rapidly become familiar with their usage. For example, when I first began studying the book[44] from which the following extract is taken, I struggled to make any sense of what was to me unintelligible text. Over time however, due to a great deal of persistence, I became quite comfortable with reading the text and whilst not fully understanding, was able to glean information sufficient for my purposes.

Here is that text:

> 16, 2 *A* (incorrectly) מֵיָּה, the rest have מִלָּה. 19, 8 B., *A* and *C* הָעִיִּין, *B* הָעִיּוּן, *D* הָעִיּנוּ. The subject being מָן *D* is of course a mistake. 20, 9 *A* (incorrectly) בְּשָׂרָן; the rest have יְשֻׁרִין. *A*, *B* and *D* are incorrect in having הָרִין 20, 13 since שִׁיבוֹתָך is feminine. B. and *C* read הִיא. *B* is incorrect in reading הִיא 19, 13 since אֹהֶל is masculine. Rest have הוּא. The Hebrew in each case is זֶה and this may perhaps account for the error of *A*, *B* and *D* in 20, 13. 21, 30 B. and *C* הָוֵא, *A*, *B* and *D* הַוִּין. Since בֵּיהֶא is given by L. as of common gender either reading may be justified; but *B*'s reading הָוֵא in 24, 58 and *A*'s הַוָּא in 25, 30 are both evidently incorrect. The fact of בֵּיהֶא being of common gender may again justify both readings in 26, 20 in which verse *A* reads מֵיָּה whilst B., *B*, *C* and *D* give מִלָּה. 27, 42 *A* and *C* (incorrectly) הָיֵא, B., *B* and *D* הֵא. 29, 3 B. לְאַשְׁקִין, which L. also reads. MSS. קְשַׁלָּן. Pathšegen also as B. who gives the proper gender, the subject being the shepherds. 30, 37 *A* and *D* הָעִיבִין, *B*, *C* and B. טִישׁוּבֵן. So also L. and Pathšegen, but since חוּטְרִין is of common gender we may take either reading as correct. 30, 40 all the MSS. have גַאֲרִיבֵין which is the proper reading, B. has עֲרִיבִין which is certainly incorrect. 30, 43 *A* סַגְרִיאָן. B., *B*. *C* and

The symbols on this page represent data which, when combined with other data found in the glossary, transform to information, but only to a reader prepared to exert the necessary effort. The scientific observer should ask: How does that happen in real-time and what are the causal factors? This example bears on such questions as: do we have free will; does the brain drive the mind or the mind, the brain;

and whether organic properties alone can account for such complex data processing? What was it about me that resulted in the necessary persistence to derive some meaning from the text? Why did I not cast the book aside?

Quantification of Information

Another definition that I have encountered is this: Information theory is a branch of applied mathematics, electrical engineering, and computer science involving the quantification of information. That is all well and good so far as it goes, but quantity is not the same as quality, and from a cognitive perspective, quantity does not convey meaning except in a context external to the data itself. A good author could write this book in half the number of words: the informational content would remain the same (or improve) but the quantification would be halved. A garrulous or loquacious person will normally convey less information than a speaker who is both precise and concise. In a cognitive sense, we evaluate information on its value and relevance, not on its quantification. More importantly, there is no necessary relationship between quantity and *humbug*.

Quantification alone can tell us nothing about the meaning of the information, how it is derived, or from an evolutionary perspective, how it is originated before there were cognitive beings that could make sense of it.

Communication

No doubt it has occurred to you that data transfer does occur in the absence of cognitive beings such as humans. Predators detect prey by sight or smell, dolphins and bats use echo-location, some creatures sense small movements and vibrations and so on: data transfer does occur. Scientists have reached a level of understanding in this domain but there is one important distinction: such messages are generally unambiguous because the coding systems are fixed by the laws of physics and chemistry. When a hyena smells a dead carcass, the communication can speak of nothing other than what it does; it cannot, for example, convey information on the height of the nearest tree, or whether the princess in a faraway land has given birth. Using the term loosely, we can say that the communications protocol is constrained by the physical properties of the source and the medium through which it passes.

Animals do have other means of communication such as leaving a scent, behaving in certain ways, changing colour, or making sounds, but to the best of our knowledge, all such communication is relevant to their current environment and circumstances. We have no evidence that animals discuss the weather, their ailments, the development of their offspring, the pros and cons of moving to a new location, whether they should craft an autobiography, or what they will be doing tomorrow.

Human communication is entirely different. Whilst we share many of the faculties of lower order animals, even to a lesser extent, these animals do not share our higher order faculties in terms of information, knowledge, and communication even though we share the same physical properties. Many articles assert that the genomes of humans and chimpanzees differ by only 4%, though I have reason to believe that such estimates are outdated. Nevertheless, irrespective of that particular difference, humans, chimpanzees, elephants and whales have 100% commonality in one fundamental respect - they are all composed of the same chemicals. They may be arranged differently, but all suffer from the same constraint - chemicals can be nothing other than what they are and cannot rearrange themselves based on an abstraction of which they cannot be aware.

CHAPTER 5

Data Processing Principles

"Computers are good at following instructions, but not at reading your mind."
~ Donald Knuth ~

Modern computer users are largely unaware of what occurs behind the user interface, and even modern programmers have little need to know much of the detail at the system level. My first programming language was at the machine level, ones and zeroes, later moving on to Assembler, Cobol, RPG, PL/1, and subsequent high-level languages. IT practitioners in the early 1970's needed to write their own data storage algorithms; understand cylinder and head addressing techniques to optimise performance; be able to read core dumps in hexadecimal or octal; and perform many other tasks that are now *black-boxed* for them in the interests of productivity. This is as it should be, but I want to review some basics to build an analogy for understanding our cognitive abilities, at least in part: much of what occurs in the human mind is beyond analogy with digital data processing.

Storing Data for Input

Firstly, we should understand that the storage of data is independent of the storage medium – the same message can be stored in many different ways. Secondly, because it is not data but symbols that are stored, a method of encoding and decoding must be devised and employed. Thirdly, like storage, the content of a message is independent of the method of transmission, and is entirely dependent upon an intelligently devised coding system known to both the sender and receiver. Speech and writing can convey the same messages, or different messages, as can smoke signals (or so I have been told). A message can be written in the sand with a stick, or sticks and stones can be arranged on the sand to form the same message. In short, there is a large variety of physical methods for storing and transmitting the same data or a single message, demonstrating that the data or message itself is conceptual, not physical.

In the early days of EDP (Electronic Data Processing), a far more explicit term to my mind, data was input using punched cards. Holes were punched in the card in a particular pattern, but the pattern was independent of the storage medium (card) and varied depending on the manufacturer of the computer system. Over time, IBM for example modified both the card size and the pattern such that early card readers could make no sense of cards punched on later machines. Each hole on a given card was the same in shape and size, the placement on the card being necessary to complete the symbol, but the symbol itself could tell you nothing about what it was meant to represent.

The point to note is that the arrangement was based on a human-devised coding system, and only recipients, machine or human, educated on that system could understand the data content. With experience, one could directly read the punched card from the holes themselves, as I have done, but first I needed to know the coding system.

Another development using the same IBM 80 column card format was known as "mark sense". Instead of punching holes in the card, one used a graphite pencil of other black marker to mark the card in predefined places. The card reader would pass the mark positions to the central processor which would then interpret them based on the coding system known to the host program. Again, the individual marks, as symbols, conveyed no meaning to the storage medium, and not even to humans without the pre-printed instructions on the card. The point here is that the same storage medium, an IBM 80 column card, could be used with different methods of recording the symbols, and with different coding systems, yet still intended to represent exactly the same data. That which was represented was independent of the method and format of how it was represented, and clearly, neither the storage medium itself nor the symbol methodology had any understanding of the data being stored.

Now you may think this rather trivial and obvious: of course inanimate materials are not sentient and are incapable of understanding data or information. Hold that thought.

Reading a punched card, it was often difficult to interpret the data punched therein without knowing the application to which the data belonged. The symbols were contiguous with no gaps, 80 symbols in 80 columns, and so often consisted of just numbers. You might see a string that looked like a date, but unless you knew that it was a date and you knew the date format (code), you would just be guessing. For example, dates were coded as YYDDD, DDMM, MMDD, DDMMYY, DDMMYYYY, etc, so 750441234 could be year 75, day 44, or it could be card 75 for month 04, and so on. Consequently, it was imperative that the appropriate computer program was loaded before the data in the form of a deck of cards was fed to the system. In essence, the processor had to pre-know both the coding system and the data formatting used in the cards.

Storing Data for Later Use

When processing was completed, the system would then store data onto media such as punched card, magnetic stripe card, magnetic tape, hard disk drive, paper tape, or paper reports via a printer using a physical method appropriate to the media type. Depending on the media, the symbols were either human-interpretable or not, but again could not be interpreted by the medium itself. You may consider that I am stressing this point and you would be right: I am and for very good reason.

The storage of human interpretable data, such as words written on paper, is managed by humans, but digitally encoded data on a hard disk in particular is stored using algorithms formulated by intelligent design. Imagine if you will an abundance of 1's and 0's scattered around on a 500Gb disk in locations and patterns known only to the computer operating system. Fiction stories tell of forensic analysts analysing computer hard drives to discern the contents, but what the stories do not relate is that first the analyst must know, even by trial and error, what computer system converted the data to symbols and using which method. The ones and zeroes are meaningless in themselves.

My technical knowledge is somewhat out of date but the principles remain true and are instructive. Between the symbols used to store data and a human interpretable form are multiple layers of coding. Symbols are stored in magnetic regions with the polarity, N-S or S-N, being sensed whilst reading that region. Each region equates to a *bit*, representing either a 1 or a 0. As you probably know, 8 bits make up a byte but after that it becomes complicated. Without going into detail, there were (are) two encoding protocols: ASCII (American Standard Code for Information Interchange) used by most computer manufacturers and EBCDIC (Extended Binary Coded Decimal Interchange Code) favoured by IBM. At an intermediate level were Octal and Hexadecimal but one only saw these in core dumps, which of course we could read as fluently as any other language. So the path from a magnetic region on a disk to an interpretable letter like an A or B involved considerable navigation based on complex coding systems developed by intelligent agents, although in those early days we hesitated over the word "intelligent".

Another issue, which should be obvious, is that having stored these binary bits in magnetic regions, the system had to know where to find them again, and find them in context: no use looking up purchasing data if you were running that week's payroll. Thus the next level of complexity is a method of addressing and indexing so that when needed, the data could be quickly and accurately retrieved. But note here also that the computer system had to "know" where the indices themselves were stored and how to decode the data to convert it into a search algorithm. Even more, and this is a point which I will stress later on, before the computer or any form of analogue or digital computing device can begin computing, it has to "know" that it can "know" what to do. I acknowledge that this point may seem obscure, but I will hopefully clarify this a little further on. In a digital computer, the first level of knowledge occurs in some bootstrap procedure which gets the show going by invoking the operating system which amongst other tasks, validates its memory and checks out its various peripheral devices whilst invoking start-up programs to awaits further instruction. When the user invokes a program, the computer has to know that such a program exists in memory and where to find it, and the executed program has to know which files to use and where to find them. I doubt that the human brain works in quite the same way, but I am in no doubt that the same functional requirements must be met.

Continuing with storage media, in earlier tape-based storage systems the program would hunt through the file sequentially but the data had to be organised sequentially when written to the tape, and the system had to foreknow that sequence: employee number followed by department or department followed by employee number?

Now here is a very important fact, one if not true would have prevented the development of Information Technology and deprived us of technical progress in numerous areas: data symbols stored on physical media cannot (and should not) self-organise. We might consider that in the context of the brain: if the organic matter self-organises, who or what does it tell about the new organisation such that the re-organised data can be found again?

Possible vs Probable

Scientists will tell you that molecules can self-organise and this is true, but to a limited extent only. Before we get into that, let us step back for a moment to examine how scientists will often claim possibilities for materials without justifying how the properties of the material would allow that possibility. Writing on the subject of miracles in *The blind watchmaker*, Richard Dawkins gave an example of how a marble statue might wave a hand by sheer coincidence; here is what he had to say:

"A miracle is something that happens, but which is exceedingly surprising. If a marble statue of the Virgin Mary suddenly waved its hand at us we should treat it as a miracle, because all our experience and knowledge tells us that marble doesn't behave like that."

He continued:

"In the case of the marble statue, molecules in solid marble are continually jostling against one another in random directions. The jostlings of the different molecules cancel one another out, so that the whole hand of the statue stays still. But if, by sheer coincidence, all the molecules just happened to move in the same direction at the same moment the hand would move back. In this way it is *possible* for a marble statue to wave at us. It could happen. The odds against such a coincidence are unimaginably great but they are not incalculably great. A physicist colleague has kindly calculated them for me. The number is so large that the entire age of the universe so far is too short a time to write out all the noughts! It is theoretically possible for a cow to jump over the moon with something like the same improbability. The conclusion to this part of the argument is that we can calculate our way into regions of miraculous improbability far greater than we can imagine as plausible."[45]

This all sounds wonderful stuff if one knows nothing of materials science, but fortunately for us an acknowledged expert in this discipline, Professor Edgar Andrews, is on hand to expose Dawkins' error. He points out that while molecules in gases may exhibit random movement, the same is not true of crystalline solids like marble. You can read his full explanation in his referenced book, but my point is that many scientists appear to believe that you can posit the probability of an event without first verifying that the event is even possible within the laws of science as we know them. The irony of Dawkins' example is that if such an event were observed to occur, the best explanation would be that it truly was a miracle! Professor Andrews concluded: *"The idea that the internal motion of atoms in a lump of crystalline rock could somehow cause that lump to move from here to there is scientifically ridiculous. Harry Potter it may be, but science it is not."*[46]

Self-Organisation

Applying this example to the subject of self-organisation, we should understand that self-organisation such as happens with certain materials can only occur based on the intrinsic properties of the materials themselves. We could hypothesise that the chemicals listed in the Periodic Table of Elements could self-organise based on their atomic number (number of protons), such being intrinsic to the elements themselves, but they could not self-organise based on their scientific names because the names are abstractions *unknown* (not intrinsic) to the elements.

From an informational perspective, the data represented by stored symbols is an abstraction and bears no natural relationship to the properties of the material itself. The symbols in whatever form, magnetic regions on a disk or electro-chemical gradients in the brain, cannot self-organise based on concepts unknown to them: for them to meaningfully represent an abstraction, their initial organisation must be established by the same agency that seeks to derive meaning from it, just as occurs with coding systems. It should be evident that in the case of abstractions, the agency causing the organisation must be external to that which is being organised.

Self-organisation alone cannot represent information that is an abstraction of that organisation. If we use the letters of the alphabet as a simplistic example, stored letters could only self-organise like-for-like: all the A's together, all the B's, etc, but not in sequence, for both "letter" and "sequence" are concepts *unknown* (not intrinsic) to the symbols themselves. What we know as an "alphabet" is a concept which provides the context for the sequence with which we are familiar, and as the letter symbols *know* neither the concept nor the context, any organisation of collections of like letters would be random. Another possibility is that the letters are stored in the sequence of arrival but that is the antithesis of self-organisation. Irrespective of the method of storage, something external is required to organise it.

Fortunately for us, stored symbols once organised by an external, intelligent agent, usually stay organised, allowing us to retrieve them in the sense intended. Sadly though, this appears to be not always true in the human brain. A little later, we will further investigate the issue of whether or not organic brain matter can self-organise.

For a practical example of the natural self-organisation of crystalline structures, see here[47]. These massive structures demonstrate what is possible within the intrinsic properties of the material, but they could not, for example, code for the latitude and longitude of their location, nor for the country in which the caves can be found, these being abstractions external to properties of the crystals.

Data Ancestors

A fundamental of *epistemology*, the discipline concerned with the nature, sources, and limits of knowledge, is that all knowledge is based on prior knowledge, which must surely lead an enquirer to question where the very first knowledge originated. We shall come back to that. For now I will introduce the notion of *data ancestors*: that long string of prerequisites which are employed in any intellectual activity. It is said that a good investigator will continue to ask "why" until he arrives at the first cause; a researcher of knowledge should continue to ask "how do you know that" until he arrives at the very first element of knowledge. In digital information processing, the same investigations can occur with precision because the structure of the data is predetermined by an intelligent agent, but we should not lose sight of the fact that it took an intelligent agent to not only organise the data, but to interpret it in context. This latter point evidences than even the largest and most complex of computer storage is still inadequate to comprehend its contents, for the definitions of the foundational concepts employed are external to the computer storage itself. Even if those definitions are also stored in that same computer, another application of external intelligence is required to provide the correlation and interpretation – again, *turtles all the way down*.

Let us now consider the structure of data ancestors.

Concepts and Context

In the early days of data processing, practitioners identified three file types: *master*, *application*, and *transaction*. Fundamentally, master files provided concepts; application files provided context; and transaction files provided activity data. As more than one computer operator learned as they loaded stacks of cards, processing the wrong transaction file against the pre-loaded master or application file failed to result in useful information, giving rise to the acronym GIGO: Garbage In Garbage Out.

The master file provided details of the entity (concept): e.g. employee, inventory item, customer, supplier, and so on. The application file provided details of the activity type (context): e.g. payroll, stocktake, sales orders, purchase orders, etc. The transaction file provided details of the activity itself. In the days of sequential and indexed-sequential file processing, IBM at one time conducted a survey to determine the most common cause of programming errors. If my memory serves me correctly, 80% of errors were the result of incorrect file matching. This is instructive in the context of cognitive information processing: if intelligent computer programmers regularly made logic errors concerning the correlation of data sources, what is the probability of even more complicated logic arising through undirected organic evolution?

The master, application and transaction files also contained concepts beyond those immediately obvious, leading to the requirement for a *data dictionary*: a precise definition of every data field used by the information processing systems. Errors in programming were commonly caused by inadequate or absent definitions. In later systems such as Business Intelligence, entirely wrong results had similar causes; for example, *Delivery Date* not adequately identified as *promised* or *actual*. Though data is now stored in random access data bases, the fundamental principles do still hold.

Knowledge is acquired through information, and information is dependent upon a structured chain or network of data ancestors which provide the concepts and context, prerequisites for accurate processing of any new data transactions however acquired.

Validation and Verification

"If debugging is the process of removing software bugs, then programming must be the process of putting them in."
~ Edsger Dijkstra ~

Program and system testing remains one of the most problematic aspects of modern software, especially as complexity increases. New software patches and releases are mostly related to fixing problems rather than adding capability. In the military aviation domain, and to a lesser extent the civilian aviation industry, computer software represents an ever increasing cost component and a primary reason for project delays. A Navy helicopter program in Australia was eventually cancelled because after ten years, the integration of numerous software applications had still not been successfully achieved. In the current F-35 Lighting II project in the USA, full software integration will not be achieved until long after FOC (Final Operational Capability) has been accepted, the compromise having to be made to forestall cancellation of the program. I mention these cases to evidence the complexity of poly-functional software systems, and would offer that the human brain has capabilities far beyond any software system yet devised.

Program design is ever a trade-off between straight-line coding which offers simplicity in understanding but increased overall size and restricted functionality, with modular coding whereby a particular code module can be reused in multiple functions reducing size but increasing complexity. A disadvantage of poly-functional modules is poly-dependency: that is, a change in a single module will result in multiple downstream changes, often unintended and consequently adverse.

An example from aviation is when an aircraft manufacturer added code to override a pilot's attempt to retard the throttles during take-off - a critical phase of flight. This action was taken based on experience in the USA and on the advice of the FAA in that country. In Sweden one winter, clear ice breaking off the wings of a DC-9 caused both engines to surge and the pilots reacted correctly by temporarily retarding the throttles to clear the surge. The software overrode the pilots' actions, the increased thrust caused both engines to explode and disintegrate, and the aircraft inevitably crashed. One could offer that the software designers failed to properly understand poly-dependency and evaluate all possible scenarios even if they could foresee all such possibilities. In the context of undirected evolution, poly-dependency would likely lead to adverse rather than fortuitous consequences more often than not.

My point here is that for any information processing system to be useful in providing valid data and true knowledge, there needs to be a validation process which is external to the system being validated. Further, the validation process itself needs to have been validated by a prior process, itself having been validated. The obvious question arises: How does one get started? The implications for the over-arching narrative of evolution are significant.

Coding and Decoding

All electronic computer and communication systems require predefined coding systems which are understood by all nodes. This aspect of the technology is perhaps the primary reason why interoperability across systems from different manufacturers remains a source of frustration even today. In a very general sense, all communications, whether verbal, written, visual, audible, or tactile rely on systems of coding. The very words you are reading here are in a code comprising letters of the alphabet, vocabulary, grammar, and punctuation. Without going into the subject in depth, practically all codes are multi-level.

It is axiomatic that the definition of a code must precede its usage, otherwise the coded symbols are no better than *humbug* (noise). My point here is that all known information processing systems involve a hierarchy of codes, and that these codes need to be predefined to all nodes in the process. This is particular importance when we later discuss cognitive dependent senses such as sight.

Can Coding Systems Evolve?

A computer processor operates via switches metaphorically set to either "on" or "off". A system with 8 switches (2^8) provides 256 options; 16 switches (2^{16}) 65,536; 32 switches (2^{32}) 4,294,967,296; and 64 switches (2^{64}) the very massive 18 trillion. This feature provides the terminology such as 32-bit and 64-bit computers: it refers to the number of separate addresses than can be accessed in memory. In the early days of expensive iron-core memory, 8 or 16-bit addressing was adequate, but with the development of the silicon chip and techniques for dissipating heat, larger memory became viable thus requiring greater addressing capability and the development of 32 then 64-bit computers. The relevance of this is found in most users' experience: software versions that are neither forward nor backward compatible. The issue is that as coding systems change or evolve, the processing and storage systems cannot simply evolve in an undirected fashion: they must be intelligently converted. Let us look at some practical examples.

Computer coding systems are multi-layered. The binary code (1's and 0's) of computers translates to a coded language such as Octal and then ASCII, or Hexadecimal and then EBCDIC, and then to a common human language such as English or French. Computer scientists long ago established these separate coding structures for reasons not relevant here. The point to note is that in general, you cannot go from Octal to EBCDIC or from Hexadecimal to ASCII:

special intermediate conversion routines are needed. The problem is that once coding systems are established, particularly multi-layered systems, any sequence of symbols which does not conform to the pattern cannot be processed without an intelligent conversion process.

Slightly off-topic, but consider the four chemicals in DNA which are referred to as A, C, G, and T. Not that long ago, scientists expressed surprise in finding that the DNA sequences code not just for proteins, but for the processing of these proteins. In other words, there is not just one but two or more "languages" written into our genome. If an evolving genome started with just two chemicals, say A and C, downstream processes could only recognise combinations of these two. If a third chemical G arose, there would be no system that could utilise it and more probably, its occurrence would interfere in a deleterious way. Quite simply, you cannot go from a 2 letter code to a 3 letter code without re-issuing the code book, a task quite beyond undirected biological evolution.

The Code Book Enigma

I will use an example similar to DNA because it is much easier to illustrate the problem using a system comprising just four symbols, in this case W, X, Y, and Z. I am avoiding ACGT simply so that you are not distracted by your knowledge of that particular science. Our coding system uses these 4 letters in groups of 3. If I sent you the message "XYW ZWZ YYX WXY" you would have no idea of what it means: it could be a structured sequence or a random arrangement, for the letters are just symbols which are individually meaningless until intelligently arranged in particular groups or sequences of which both the sender and receiver are aware. To be useful, we would need to formalise the coding sequences in a code book: that way the sender can encode the message and someone with the same version of the code book can decode the message and communication is achieved. Note that if the sender and receiver are using different versions of the code book, communication is compromised.

This brings us to a vital concept: meta-data (or data about data).

There is a foundational axiom that underpins all science and intellectual disciplines—*nothing can explain itself.* The explanation of any phenomenon will always lie outside itself, and this applies equally to any coding system: it cannot explain itself. You may recall the breakthrough achieved by archaeologists in deciphering Egyptian hieroglyphs when they found the Rosetta Stone.

In our example, the code book provides the meta-data about the data in the encoded message. Any language, particularly one limited to just four letters, requires a code book to both compose and decipher the meaning. Every time there is a new rearrangement of the letters, or new letters are added to (or deleted from) an existing string, the code book has to be updated. From a chronological perspective, for a change to be functional, the code explanation or definition must precede, not follow, any new arrangement of letters in a message. Rearrangements that occur independently of the code book cannot be understood by downstream processes. Logically, the code book is the controlling mechanism, not the random rearrangements of coding sequences. First the pre-arranged code sequence, then its implementation. In other words, it is a top-down process, not the bottom-up process that evolutionists such as Richard Dawkins assert.

Now it matters not whether you apply this to the evolution of the genome or to the development of our senses, the same principle holds: ALL data must be preceded by meta-data to be comprehensible as information. In general, messages or other forms of communication can be considered as transactions which require a conceptual and contextual framework to provide meaning.

An Example of Intellect

You may have seen this before, but I thought it worth sharing as it adds another dimension to the complexity of how we humans intuitively process information.

One manager let employees know how valuable they are with the following memo:

"You Arx A Kxy Pxrson"

"Xvxn though my typxwritxr is an old modxl, it works vxry wxll xxcxpt for onx kxy. You would think that with all thx othxr kxys functioning propxrly, onx kxy not working would hardly bx noticxd; but just onx kxy out of whack sxxms to ruin thx wholx xffort.

You may say to yoursxlf, "Wxll I'm only onx pxrson. No onx will noticx if I don't do my bxst." But it doxs makx a diffxrxncx bxcausx to bx xffxctivx, an organization nxxds activx participation by xvxry onx to thx bxst of his or hxr ability. So thx nxxt timx you think you arx not important, rxmxmbxr my old typxwritxr.

You arx a kxy pxrson."

Notice how even when certain symbols are replaced, it is possible to understand the message by subconscious substitution, further evidence that abstracted meaning is independent of the intrinsic properties of a stored (on paper) symbols. I will repeat that for it is an important point in relation to cognitive processing: without any training whatsoever, and without any pre-arrangement or warning, we can intuitively correct errors when the written code does not accurately follow the code book. How can an undirected process of evolution account for such an instant error detection and correction mechanism?

What particularly fascinated me when I first encountered this little gem was how I was able to immediately start reading it despite the spelling errors. As a former computer programmer, I wondered what algorithms would be needed in software to make sense of this on the first pass. I have some familiarity with software that "reads aloud" from written text, but this passage would certainly present a challenge.

Data to Information

Symbols stored on physical media are not data until *decoded*, and similarly data is not information until processed within a referential framework of *concepts* which provide *context*. Let me provide a simple scenario to demonstrate this point.

You discover a piece of paper and on it you notice these symbols:

• ••• •—••

The first question that you would likely ask yourself is this: Is this a random string of symbols or is there a pattern to it – is it meant to be information or just doodling? Incidentally, this is the same thought process used by SETI researchers in attempting to identify extra-terrestrial intelligence. If the symbols are not random, then we can assume that there is an intelligence behind the structure and a purpose behind a communication, the layer described by Dr Werner Gitt as *Pragmatics*[48]. In passing, note that I am deliberately avoiding the excellent work by Shannon, Gitt and others so as to not distract from the main theme. However, I wanted to point out that *purpose* is a *concept* which needs to have had an origin of its own. Now note the structure of the symbols: one round thing by itself, three round things together, and so forth; even the spaces between them are symbols like punctuation (another concept).

Let us assume that you are aware of Morse Code, another concept which provides the context for understanding the symbols. Perhaps because you have encountered its usage in SOS messages, you can now transliterate the message as *dit, dit dit dit, dit dah dit dit*, just as the SOS is three short, three long, three short. You are making progress, but remember that chain of data ancestors that you needed to get to this point? Off you go to the library or the internet and find that these three symbols represent the English letters, ESL. So far we have dealt with the *concepts* of purpose, communication, symbols, patterns, formal code sets, and translation; each *concept* refined the *context* for what followed. Your next problem is to understand the context of the abbreviation: is it related to chemistry, physics, history, archaeology, traffic management, politics, or a board game, but in asking that question, you must first be aware that such domains of information exist (your search space): again an issue of concepts to provide context. If I told you that it related to *axsebology*, you would be none the wiser (because I just made it up). The actual domain is air navigation, and the message is the identification code broadcast from an NDB (non-directional beacon). You now know more than you did to start with because you have been informed of new concepts which further refine the context of the message, but you are still clueless as to where that NDB may be located. You can find a table of descriptions and locations, and you already knew that you could, and eventually discover that the NDB is at the Royal Australian Air Force Base, East Sale, Victoria, Australia.

For such a very simple message consisting of just two symbols, or three counting the spaces, there have been a multitude of prerequisite concepts (data ancestors) needed to provide the context for you to refine your search space to just those concepts which provide the correct solution. Note that all of those concepts had to be acquired from sources external to you, whether you already pre-knew them or just then acquired them. What we are seeing, to use the term coined by William Dembski, is *Complex Specified Information* (CSI). There is an extensive network of information that could have been explored to find the solution, but only a *specific set* would get you there successfully. Most data points encountered in the process have multiple links: in fact, information is a complex network of data points that need to be correlated in a specific way to provide the right answer, but then the question arises: how do you know that you have arrived at the right answer? Just as in Quality Control, there needs to be a process of validating the outcome of the upstream process. As we know from experience, people can evaluate the same data yet arrive at different conclusions. The reasons are multifarious, ranging from our worldview to professionally acquired knowledge and experience, but the key point is that for any successful transfer of information, there must be a rigorous method of validation: randomness will simply not do.

Summary

In this brief review, I have attempted to outline some fundamentals which need to be present in any information processing system. In researching this study, I at one time attempted to start with a single, simple concept like a "menu" and document the data ancestors which I unconsciously used in understanding what a menu is and how it is used. I used my experience as a systems and database designer to structure the concepts and contexts in a way that would lead me from bottom to top, or even top to bottom. After some three hours and many pages, I had a good appreciation of the depth of my knowledge and the utter futility of trying to document it.

Later, I will use the following fundamentals in relation to human cognitive functions:

- symbols
- coding systems
- data ancestors
- concepts
- contexts
- search space
- validation
- purpose
- storage and retrieval

There is no suggestion that the *mechanisms* or *architecture* of computer systems must be present in the human mind and/or brain to accomplish cognitive information storage and processing. I do assert, however, that the functional requirements are the same, irrespective of the method of implementation.

CHAPTER 6

Information Processing in the Mind

"No matter how closely you examine the water, glucose, and electrolyte salts in the human brain, you can't find the point where these molecules became conscious."
~ Deepak Chopra ~

In this brief chapter, I will allude to our sense of sight and then discuss it in more detail in subsequent chapters.

Let me declare my conviction that despite being beyond my understanding, the mind is the immaterial controller of the non-autonomous functions of the brain. In other words, the brain is organic but the mind is not. I have read numerous academic texts on this subject but find it impossible in the context of cognitive information processing to treat the two as one. Though the centrality of my argument here is with the overarching narrative of evolution, and for the sake of simplicity I would prefer to consider the mind and brain as synonymous, I cannot manage to discuss the mind as an organ of physical properties alone.

The brain is connected to numerous data input sources but here I want to consider just the senses which may contribute to the development of knowledge in the intellectual sense. I will also disregard structural complexities such as the visual cortex (for now) and consider the brain as a single organ.

In the context of sight, data arrives in a coded format appropriate to the method of transmission. For example, photons of light arriving at the eye are transformed into electro-chemical signals transmitted by the optic nerve. This reveals yet another level of complexity which I shall ignore for the sake of simplicity. These data symbols are placed in physical storage. A relevant lesson from the information sciences is that information is independent of the medium in which it is stored, and any organisation must be implemented by something external to the medium itself. Here we should revisit a related axiom: nothing can explain itself. This poses a problem for the evolutionist: how can data symbols self-organise in such a way that they can represent concepts of which they are unaware?

Emergence and Artificial Intelligence

The evolutionists' answer is that information and the intelligence to process it are *emergent properties* of the brain. Of course, no-one has yet to propose a plausible let alone proven mechanism for such emergence. In physics there are discussions on *Self-Organised Criticality* (SOC), no doubt an important field in statistical physics related to dynamical systems, but again this field only deals with intrinsic properties, not abstractions. Let us spend a moment on this concept of emergence; Karl Popper noted: "The idea of 'creative' or 'emergent' evolution ... is very simple, if somewhat vague. It refers to the fact that in the course of evolution new things and events occur, with unexpected and indeed unpredictable properties; things and events are new, more or less in the sense in which a great work of art may be described as new."[49] Clearly, emergence is not an observed process, but presumption of a process within the presupposition of the evolution paradigm: *somewhat vague* is an understatement! Students of logic should recognise this as the fallacy of begging the question.

In the evolution context, the proposition of emergent properties is hardly better than the earlier scientific beliefs in spontaneous generation, e.g. that maggots arise from dead flesh: it assumes spontaneous generation of the initial state from which more complex properties can emerge. I have read a number of scientific papers on emergence but none so far attempt to explain the origin of these initial sub-processes that can give rise to higher level processes. The primary purpose of such scientific research is to be able to model the brain leading to artificial intelligence, but it is illogical to propose that the model can explain its own origins, particularly when it is based on an unproven hypothesis. In the evolution context, the explanations that I have read simply beg the question; for example, this quote from the University of Maryland:

> "The first step to applying rule abstraction to the brain and mind, as with any complex system, is by **declaring the obvious**: the cognitive powers of the mind and brain result from the physiology's emergent properties. This statement represents the initial state of the hierarchy."[50] [emphasis mine]

I have no issue with the methodologies and goals of these researchers as they are directed at understanding how the mind works, not how it arose. I do, however, contend with their assumption that cognitive powers of the mind result from physiology, as the latter is a sub-discipline of biology which deals only with the material. Studies of cognition are in the domain of metaphysics, not the physics or chemistry. Simply declaring it to be obvious is not science - it is philosophy, and poor philosophy at that: physiology deals only with intrinsic properties, not abstractions.

The quote above highlights what is to my mind, a blind spot in the perspective of the researchers. They note the necessity for *rule abstraction* in cognitive processing, yet seem to gloss over it as if it were a minor issue - in truth, it is THE issue. Abstraction is what differentiates cognitive processing from genetic, organic, or other type of material information processing. The latter can only occur using their intrinsic properties, but by definition, a logical process is required to effect the abstraction: there is no logic intrinsic to organic materials, only properties.

In the cognitive sense, *abstraction* refers to ideas rather than real events. Chemistry deals only with real events, not with ideas, and nothing in the physical sciences offers a solution that bridges the chasm between events and ideas. We know that ideas exist, they are the foundations of scientific research, yet strangely, science proceeds as if the conceptual created itself out of the material which has none of the properties of the conceptual.

The other issue requiring examination is that of *rule abstraction*. Can rules, *aka* logic, arise from undirected organic processes? Certainly there are rules (laws) regulating chemistry and physics, but note that no-one knows where such rules come from. Even so, such rules are simple in comparison to the rules of logic, particularly where such logic involves the interpretation and correlation of the abstract or conceptual. The laws of science are constrained by the properties of space, time, energy, and matter – not so the rules of logic.

For a property of the mind to be emergent, the mind must first exist, or as these researchers assume: there has to be an initial hierarchical state from which more complex hierarchies can emerge. From the words of the researchers themselves, the initial hierarchy is not explained by emergence and neither is any initial state of the mind.

As with so much of science, from cosmology to evolutionary biology, of necessity research must start in the middle and work its way outward, backward to origins and forward based on what has been discovered. However, the danger lies in placing too much faith in presuppositions based on the current paradigm despite the obvious contradictory reality. Until researchers accept that the gulf between the physical and conceptual must be bridged before progress on artificial intelligence can be made, they have condemned themselves to an exercise in futility.

I believe it safe to contend that machines achieving anything like the consciousness of humans resides comfortably in the domain of science fiction, but bears no resemblance to reality. Fifty or sixty years on from researchers starting work in this field, there is no discernible progress - machines can still only do what they are programmed to do. You may remember *Deep Blue*, a powerful computer that was taught to play chess and managed to defeat human Grand Masters. Sadly for Deep Blue, all it could do was to play chess - faced with a game of checkers or mahjong, it would fail to compute. Similarly, Deep Blue could not decide for itself whether it wanted to play, or whether altruistically, it wanted to play badly to let its opponent win.

Even those machines which are taught to learn can still only learn using the rules and data provided - they have limited ability to make up their own rules and cannot arbitrarily refuse to invoke a rule - rules based systems lack the concept of volition. Similarly, such machines cannot make "mental leaps" or behave intuitively, and will never be able to do so until somewhat figures out how humans do it and programs it into the machine. No doubt you may object, quoting heuristics and how computers can learn from data and a history of events, but what they cannot do is invent data or concepts which have not been introduced by an external agent.

Let me give an example.

I recently read a novel based on medical technology where an application was developed for autonomous diagnosis by a smart phone, the computer having instant access to everything ever known in medical science. The application would diagnose a medical condition and propose a solution. In the case of diabetes, insulin would be released from a surgically implanted storage device. The problem arose when the application decided to overdose a terminally ill patient based on a subjective evaluation of quality of life. According to the narrative, the application worked this out for itself using its advanced heuristic logic. The story read well, but was fanciful from a technological perspective. If the medical application was devised to cure or retard the progress of illnesses, the application could not decide on termination unless such an option was provided in the logic - termination is a concept which has to be acquired from an external agent. Similarly, *quality of life*: such a concept is knowledge based on information supported by a network of data and rules necessary to arrive at any conclusion.

No matter how well computer programs are devised to "learn" from data acquired over time, any result can only be within the context of the concepts predefined to the system: nothing new can be added. Computers deal in symbols which require predefined rules to become data and even more rules to become cognitive information. Computer programs cannot conceive of new concepts, for concepts are cognitive abstractions, ideas even. Any fears of computers becoming more intelligent than humans are totally unfounded.

Returning to the quote from the University of Maryland, the statement clearly begs the question, and the assumption of the so-called "obvious" demonstrates that the researchers do not know how to get started. They are trying to build on their assumption, but not being able to validate that assumption, they cannot start to build an AI machine. It should be obvious, even to scientists, that until they actually understand how human cognitive powers work, they cannot begin to emulate them. Still, the media and the public in general continue to buy the fiction, perhaps because they have to. Once they start to doubt, the web of the overarching narrative of evolution may start to crumble

Information Storage, Assurance, and Evolution

Now back to the main theme.

Rewinding evolution to the earliest organism that had something that could be described as a brain, let us assume some form of sensory input. What survival advantage would be offered by the storage of these sensory data symbols before there was any mechanism for processing them? Let us assume that it just happened: that simply by chance, genetic mutations occurred that resulted in data signals from a surface receptor being transmitted along a nerve creating a permanent imprint in some physical manner. Then what? We need another mutation that adds functionality which organises that data in some meaningful arrangements, but we are still not there yet. At this point we need a validation process that adds certainty otherwise the organisation could represent anything at all (*humbug*) - or more likely nothing at all.

Note too that not all skin cells are sensitive and connected to nerves. In 2014, for the first time, scientists converted human skin cells into functional pain-sensing nerve cells, what is informally termed, "pain in a dish". This is another level of complexity that evolution must explain: why in the first instance the brain "understood" the sensation as pain; why it considered pain to be useful as a warning of harm to the organism; and before it came to that understanding, why that sensation contributed to evolutionary success.

Those of us who have worked in large, complex, integrated, networked computer systems know only too well the difficulties associated with quality assurance. As best as I can understand, there is not a system in the world that approaches the complexity of the brain's neural network, yet not only did that supposedly evolve organically, but so did the quality assurance processes necessary to ensure that the right answers were produced, at least some of the time.

The brain is basically a complex network of neurons which pass chemical or electrical signals to other neurons and downstream to other parts of the body. If cognitive processing is purely an organic function and the mind developed through genetic mutation, then scientists should be able to identify differences in cell structure of neurons, synapses, or axons which could account for variations in thought patterns, computational power, reasoning ability, beliefs, preferences, creativity, apparent volition, and so forth. Stating that it is the pattern of connections between the neurons which account for variations in cognitive behaviour simply begs the question: there needs to be an explanation for the organic mechanism for routing and rerouting the connections, and that mechanism had to arise through whatever process evolutionists now favour. I am not mathematician, but if even such were possible, I do wonder at the probability of a complex network self-organising in a manner from which intelligence can emerge in such a short timescale.

I will pose another problem for evolutionists: memory allocation for the storage of data symbols and a process for knowing from where to retrieve them. If there is not an intelligent agent acting as the "operating system", then it must be the data symbols themselves self-organising in a complex, poly-functional network to be able to solve multiple context-dependent problems from the same data. Not likely I would think.

CHAPTER 7

The Irreducibility of Sight

"Try describing an elephant to a blind man"

The saying above evidences our understanding that it is practically impossible to explain something to someone who has no prior knowledge of the concept, the entity, or any of its components. We might as well try to explain a *tretwopary* or a *melopricant*: where do we start? With that in mind, let us dig a little deeper.

The Inter-Relationship of Senses

I contend that our sense of touch, by itself, cannot convey cognitive information. Imagine an unfortunate man who is both totally blind and totally deaf (not just legally so): How could you teach him to "read" Braille? How could you teach him to identify and name fruits such as strawberries or raspberries, or to identify and name certain odours and develop the "nose" of a wine taster? Touch, taste, and smell work to the extent of differentiating between sensations, but without other inputs, cannot contribute to communicable information or knowledge.

We can also question the origin of concepts such as desirable versus undesirable. Was it a genetic mutation that inspired an organism to remark to itself: "That smells nice" or "that tastes delicious"? I accept that chemistry plays the major role in the identification of odours and tastes, and perhaps again it is chemistry that determines an individual's preferences, but chemistry itself cannot give rise to the conceptual. The link between the sensations of touch, taste, and smell, and the cognitive recognition of what such sensations represent, falls into the domain of the conceptual, not the physical, especially when we take the next step to communicate our responses to these sensations to other sentient beings.

Now let us look more closely at just one sense: that of sight.

The Concepts of Light and Sight

Imagine that some thousands of years ago, a mountain tribe suffered a disease or mutation such that all members became blind. Generation after generation were born blind and eventually even the legends of the elders of being able to see were lost. Everyone knew that they had these soft spots in their heads which hurt when poked, but nobody knew if they had some intended function. Over time, the very concepts of light and sight no longer existed within their tribal knowledge. As a doctor specialising in diseases and abnormalities of the eye, you realise that with particular treatment you can restore the sight of these people. Assuming that you are fluent in the local language, how would you describe what you can do for them? How would you convey the concept of sight to people to whom such an idea was beyond their understanding? You could hardly just walk up to them and exclaim: "I can restore your sight!"

My point is that this is the very same problem that primitive organisms would have faced if sight did in fact evolve organically in an undirected fashion. When the first light sensitive cell hypothetically evolved, the organism had no way of understanding that the sensation it experienced represented light signals conveying information about its environment: light and sight were concepts unknown to it.

The Training of Sight

Those familiar with the settlement of Australia by Europeans in the 19th century, and the even earlier settlements in the Americas and Africa, would have heard of the uncanny ability of the indigenous population to track people and animals. It was not so much that their visual acuity was better, but that they had learned to understand what they were seeing. It was found that this tracking ability could be learned and taught to others. In military field craft, soldiers are taught to actively look for particular visual clues and features. In my school cadet days, we undertook night "lantern stalks" (creeping up on enemy headquarters) and later in life, the lessons learned regarding discrimination of objects in low light were put to good use in orienteering at night. All of this experience demonstrates that while many people simply "see" passively, it is possible to engage the intellect and "look" actively, thus apprehending much more.

With the advent of the aeroplane came aerial photography and its application during wartime as a method of intelligence gathering. Photographic analysis was a difficult skill to acquire - many people could look at the same picture but offer differing opinions as to what they were seeing, or rather thought they were seeing.

The underlying lesson is that sight is more a function of intellect than passively receiving light signals through the eyes. Put another way, it is intelligent data processing.

Understanding Data vs Information

As discussed in an earlier chapter, data only becomes *cognitive information* when intelligently processed against a pre-loaded referential framework of conceptual and contextual data. Using this computer analogy, master files represented *conceptual* information, application files provided *context*, and *input* data was provided by transaction files.

With apologies to Claude Shannon, Werner Gitt and other notables who have contributed so much to our understanding on this subject,

I would contend that in the context of this discussion, none of these files contain *information* in the true sense: each contains *data* which only becomes usable information when intelligently correlated. I would further contend that no single transmission in any form can stand alone as information - absent of a preloaded conceptual and contextual framework in the recipient, it can only ever be a collection of meaningless symbols. This is easily demonstrated by simply setting down everything you have to know before you can read and understand these words written here, just as I earlier related concerning *menus*, or what you would have to know before reading a medical journal in a foreign language in an unfamiliar script such as Hebrew or Chinese.

Consider that you have four documents before you: one in English, one in French, one in Chinese, and one in Russian. You are told that the information content of each is the same. Consider that your only language is English (I speak Australian, purportedly a form of English). You can recognise some, or possibly all of the elements in the English version, can recognise some of the symbols in French, but have no idea about the other two. The issue is the multiple levels of coding, starting with the symbol layer. One document has symbols (letters of the alphabet) which you recognise, one has symbols which you mostly recognise, and the other two could be anything. At the next level, vocabulary and grammar, you may be conversant with all of the words in the English version, and some of the words in the French version due to their similarity. But then we have context and the difficulty associated with words having a semantic range, sometimes culturally conditioned. For example, in the English tradition, a "rubber" is used to erase mistakes, whereas in America it is used to prevent a mistake of an entirely different kind. In Australia, a "thong" is a form of casual footwear, but in America it is an item of clothing used to (partially) cover parts of the body unmentionable in polite company.

Punctuation is also significant in conveying data in the correct sense. Consider these words from the song, *I Don't Know How to Love Him*: "For I have had so many men before in so many ways he's just one more." Does the comma go after the word "before", or after the word "ways"? The difference in the information conveyed is significant.

Finally, speech communication also depends on pronunciation, which in many cases does not follow the rules and has to be taught. I am still bemused by the American pronunciation of *Kansas* versus *Arkansas*, and *buoy* verses *buoyant*, but lest you think me racist, we have even more words in Australia which still confuse me and I am criticised when I fail to use the local pronunciation.

The point to understand is that the pathway from symbols stored on physical media, whether synthetic or organic, to information, is tortuous and multi-layered, a system far beyond the capabilities of undirected organic mutation and selection. Organic material contains none of the intrinsic properties which can account for the complex organisation of the storage and communication of the elements of cognitive information.

Understanding Input Devices

We have five physical senses: sight, hearing, smell, taste, and touch, and each requires unique processes, whether physical and/or chemical. What may not be obvious from an evolution standpoint is that for the brain to process these inputs, it must first know about them (*concept*) and how to differentiate amongst them (*context*). Early computers used card readers as input devices, printers for output, and magnetic tape for storage (input & output). When new devices such as disk drives, bar code readers, plotters, etc were invented, new programs were needed to "teach" the central processor about these new senses. Even today, if you attach a new type of reader or printer to your computer, you will get the message "device not recognised" or similar, if you do not preload the appropriate software. It is axiomatic that an unknown input device cannot autonomously teach the central processor about its presence, its function, the type of data it wishes to transmit, the context of that data, or the protocol to be used. Note that I said *autonomously* - an intelligent external agent may well have preloaded the software necessary to achieve those tasks.

The same lessons apply to our five senses. If we were to hypothesise that *touch* was the first sense, how would a primitive brain come to understand that a new sensation from say *light* was not just a variation of *touch*?

Signal Processing

All communications can be studied from the perspective of signal processing, and without delving too deeply, we should consider a just few aspects of the protocols. All transmissions, be they electronic, visual, or audible are encoded using a method appropriate to the medium. Paper transmissions are encoded in language using the symbol set appropriate for that language; sound makes use of wavelength, frequency and amplitude; and light makes use of waves and particles in a way that I cannot even begin to understand. No matter, it still seems to work. The issue is that for communication to occur, both the sender and receiver must have a common understanding of the communication protocol, the symbols and their arrangement, and both must have equipment capable of encoding and decoding the signals.

Now think of the eye. It receives light signals containing data about size, shape, colour, texture, brightness, contrast, distance, movement, etc. The eye must decode these signals and re-encode them using a protocol suitable for transmission to the brain via the optic nerve. Upon receipt, the brain must store and correlate the individual pieces of data to form a mental picture, but even then, how does it know what it is seeing? Movement may be interpreted by comparing signals in a manner equivalent to frame-by-frame analysis, but that is a very intensive data processing activity: did the early eyes not comprehend movement? How did the brain learn of the concepts conveyed in the signal such as colour, shape, intensity, and texture? Evolutionists like to claim that some sight is better than no sight, but I would contend that this can only be true provided that the perceived image matches reality: what if objects approaching were perceived as receding? Ouch! What if a shadow was perceived as a solid object?

As we discussed earlier in relation to *search space*, an empty or embryonic brain contains nothing which can be used to compare the inputs signals, and thus has no cognitive sense of reality to begin with.

When the telephone was invented, the physical encode/decode mechanisms were simply the reverse of one another, allowing sound to be converted to electrical signals then reconverted back to sound. Sight has an entirely different problem because the decode mechanism in the brain is entirely different organically to the encode mechanism in the eye, and the conversion is to yet another format for storage and interpretation. These two encoding mechanisms must have developed independently, yet had to be coherent and comprehensible with no opportunity for prolonged systems testing. Again, I am not a mathematician, but the odds against two coding systems developing independently yet coherently in any timespan must argue against it ever happening.

Data Storage and Retrieval

Continuing our computer analogy, our brain is said to be the central processing unit and just as importantly, our data storage unit, reportedly with an equivalent capacity of 256 billion gigabytes (or thereabouts). In data structuring analysis, there is always a compromise to be made between storage and retrieval efficiency. The primary difference from an analysis perspective is whether to establish the correlations in the storage structure thus extending the storage time but optimising the retrieval process, or whether to optimise the storage process and later mine the data looking for the correlations. In other words, should the data be indexed for retrieval rather than just being sequentially or randomly distributed across the storage device. From our experience in data analysis, data structuring, and data mining, we know that it requires intelligence to structure data and indices for retrieval, and even greater intelligence to make sense of unstructured data.

Either way, considerable understanding of the data is required. Is it possible that undirected evolutionary processes could account for the efficient processing of information such as we experience?

Now let us apply that to the storage, retrieval, and processing of visual data. Does the brain store the data then analyse, or analyse and then store, all in real time? Going back to the supposed beginnings of sight, on what basis did the primitive brain decide where to store and how to correlate data which was at that time, just a meaningless stream of symbols? What was the source of knowledge and intelligence that provided the logic of data processing?

Correlation and Pattern Recognition

In the papers that I have read on the subject, scientists discuss the correlation of data stored at various locations in the brain. As best as I understand it, no-one has any idea of how or why that occurs. Imagine a hard drive with terabytes of data and those little bits autonomously arranging themselves into comprehensible patterns. Quite apart from aspects of materials science, you would assert such to be impossible, but that is what evolution claims, for an organic medium at least. It is possible for chemicals to self-organise based on their physical properties, but what physical properties are expressed in the brain's neural networks such that self-organisation based on conceptual abstractions would be possible? I admit to very limited knowledge here but as I understand it, the brain consists of neurons, synapses, and axons, and in each class, there is no differentiation: every neuron is like every other neuron, and so forth. Now, even if there are differences such as in electric potential or electro-chemical gradients, the differences must occur based on physical properties in a regulated manner for there to be consistency. Even then, the matter itself can have no "understanding" of what those material differences mean in terms of its external reality. Remember that what are stored are *symbols*: encoded representations of abstractions at a very primitive level.

In the case of chemical self-organisation, the conditions are preloaded in the chemical properties and thus the manner of organisation is pre-specified. When it comes to data patterns and correlation however, there are no pre-specified properties of the storage material that are relevant to the data which is represented, whether the medium be paper, silicon, or an organic equivalent. It can be demonstrated that data and information is independent of the medium in which it is stored or transmitted, and is thus not material in nature. Being immaterial, it cannot be autonomously manipulated in a regulated manner by the storage material itself, although changes to the material can corrupt the data being represented. I do that regularly whilst typing but sometimes Microsoft, an intelligently devised external agent, automatically corrects my mistakes for me, and sometimes it makes them worse. I still cannot get Microsoft Word to permanently retain my language selection as English (Australian): quite autonomously, it chooses to revert to English (US) and reports spelling mistakes which are not.

Pattern recognition and data correlation must be learned, and that requires an intelligent agent that itself is preloaded with conceptual and contextual data, and has a method of verification.

Facial Recognition

Facial recognition has become an important tool for security and it is easy for us to think, "Wow! Aren't computers smart!" The "intelligence" of facial recognition is actually an illusion: it is an algorithmic application of comparing data points and it does that very well, but what the technology cannot do is identify what *type* of face is being scanned. In 2012, Google fed 10 million images of cat faces into a very powerful computer system, designed specifically for one purpose: that an algorithm could learn from a sufficient number of examples to identify what was being seen. The experiment was partially successful but struggled with variations in size, positioning, setting and complexity. Once expanded to encompass 20,000 potential categories of object, the identification process managed just 15.8% accuracy: a huge improvement on previous efforts, but nowhere near approaching the accuracy of the human mind.

The question raised here concerns the likelihood of evolution being able to explain how facial recognition by humans is so superior to efforts to date using the best intelligence and technology, particularly when an essential component would have been absent: a method of verification that what the animal thought it was seeing matched the reality.

Irreducible Complexity

Our sense of sight has many more components than described here. The eye is a complex organ which would have taken a considerable time to evolve, the hypothesis made even more problematic by the claim that it happened numerous times in separate species (convergent evolution). Considering the eye as the input device, the system requires a reliable communications channel (the optic nerve) to convey the data to the central processing unit (the brain) via the visual cortex, itself providing a level of distributed processing. This is not the place to discuss communications protocols in detail, but very demanding criteria are required to ensure reliability and minimise data corruption. Let me offer just one fascinating insight for those not familiar with the technology. In electronic messaging, there are two basic ways of identifying a particular signal: (1) by the predefined purpose of the input device, or (2) by tagging the signal with an identifier which the receiver "understands". For evolution to be true, we can eliminate (1) as no organism can become aware of the purpose of an evolved feature. (2) is problematic because there is no mechanism for the recipient of the signal to be made aware of the nature of the signal other than by the signaling process itself which the recipient is yet to understand.

A certain amount of signal processing occurs in the eye itself; particular receptor cells have been identified in terms of function: M cells are sensitive to depth and indifferent to color; P cells are sensitive to color and shape; K cells are sensitive to color and indifferent to shape or depth. As an aside, my personal experience has me doubting that an individual cell can be sensitive to depth. I recall being taught in high school that it was our bifocal vision that allowed depth perception: somehow the vision system made a calculation based on the delta of the eye positions. Wondering how pirates with an eye patch could make their way around so well, I used to close one eye, and then the other, looking at various objects to determine if they seemed closer or further away, and concluded that what I was being taught was not likely true. Sensibly, I did not seek to refute our rather stern teacher. Piloting an approach to a runway, I would judge height above the ground by the perceived width of the runway. As an Air Traffic Controller, I could spot aircraft at varying distances in a clear sky, without any surrounding structures to act as guides, and sometimes I could judge their distances reasonably well. I have no idea of how that worked, I just know that it did. Incidentally, here is a very interesting article sourced from here[51] regarding how bees land without crashing, and further illustrates how sight is a cognitive activity more than just the passive reception of light.

> "Landing safely is a difficult aspect of flight, because the rate of approach must be reduced to near zero at touchdown.
>
> This is hard enough on horizontal surfaces, but even more challenging as inclination increases, i.e. when landing on surfaces of different orientation. Yet honey bees achieve this easily, hundreds of times per day.
>
> To the amazement of engineers who had unsuccessfully tried lasers, radars, sonars and GPS technology in striving to design autonomous landing systems for flying robots, the bees' guidance strategy is "surprisingly simple".[52] Experiments show that bees land safely by simply ensuring that the surface they are approaching expands at a constant rate within their field of vision.[53] This is a form of *optic flow* monitoring,[54] which we have noted before.[55]
>
> Mandyam Srinivasan, professor of visual neuroscience at the University of Queensland, Australia, explained:

"If you come in [to land] at a constant speed, the image [of the landing strip] appears to expand faster as you get closer. But if you keep the rate of expansion of the image constant, you automatically slow down and by the time you make contact you're moving at almost zero speed."[56]

Mathematical modelling showed that the bees' simple visual 'autopilot' technique worked on almost any type of surface—including walls and flowers—and did not need any information about airspeed or distance from destination.

"Why didn't we think of this before?" lamented Professor Srinivasan.[2] He said that robotic aircraft could soon be equipped to mimic the bee's landing strategy using a simple, lightweight video camera. The image-only landing technique could also be applied to stealth military planes (no radar or sonar for an enemy to detect) and spacecraft (landing on other planets without GPS to guide them). However, it's most doubtful that the computer required for this programming would be as tiny as a bee's brain!

It's surely self-evident that no 'guidance strategy' came about by itself."

But I digress, now back to the discussion.

The question we must ask is how an undirected process could inform the brain about these different signal types and how they are identified. The data is transmitted to different parts of the brain for parallel processing, a very efficient process but one that brings with it a whole lot of complexity. The point to note is that not only does the brain have the problem of decoding different types of messages (from the M, P, and K cells), but it has to recombine this data into a single image, a complex task of co-ordinated parallel processing.

I have been aware for some time that the image received by the eye is inverted in transmission to the brain, just as trying to use an astronomical telescope for terrestrial purposes results in inverted images.

I have wondered how the primitive organism that first developed an embryonic lens ever understood that the image was inverted, and then by evolutionary processes managed a corrective. Just recently I encountered another curiosity. My wife underwent surgery to correct a tear (pronounced *tare*) in the back of the macular of her left eye. The procedure involved draining the fluid and inserting a gas bubble. As the gas bubble was absorbed, the eye would refill with fluid and the level could be seen as a line. Over the days that this process occurred, my wife perceived that the line was moving from top to bottom, whereas in truth it was bottom up – the fluid replacing the lighter gas. Here is evidence of the inverted image, yet curiously in this instance, the brain did not understand that it was inverted, even though it knew that all other images were. I have no explanation, and I doubt that anyone does, but it does suggest that the correction of the inverted image is not an autonomous response by the brain.

But back to the main theme, finally we have the processor itself which if the evolution narrative is true, progressively evolved from practically nothing to something hugely complex. If we examine each of the components of the *sight system*, it is difficult to identify a useful function for any one of them operating independently except perhaps the brain. However, absent of any preloaded data to interpret input signals from wherever, it is no more useful than a computer without an operating system. It can be argued that the brain could have evolved independently for other functions, but the same argument could not be made for those functions pertaining to the sense of sight.

As best as I can understand, our system of sight is irreducibly complex.

Inheriting Knowledge

Let us suppose, contrary to all reason and everything that we know about how knowledge is acquired, that a primitive organism somehow began developing a sense of sight. Maybe it wandered from sunlight into shadow and after doing that several times, came to "understand" these variations in sensation as representative of its external environment. Just what it understood and how it got it right is anyone's guess, but let us assume that it happened. How is this knowledge then inherited by its offspring for further development? If the genome is the vehicle of inheritance, then sensory experience must somehow be stored therein and be progressively built upon for further cognitive development to occur.

I have no answer to that, but I do wonder.

Putting it all together

I could continue to introduce even greater complexities that are known to exist, but I believe that we have enough to draw some logical conclusions. Over the past sixty years, we have come to understand a great deal about the nature of information and how it is processed. Scientists have been working on artificial intelligence with limited success, but it would seem probable that intelligence and information can only be the offspring of a higher intelligence. Even where nature evidences patterns, those patterns are the result of inherent physical properties, but the patterns themselves cannot be externally recognised without intelligence because "pattern" is a concept which must be learned. A pattern is a form of information, but without an understanding of what is regular and irregular, it is nothing more than a series of data points. Most importantly, any pattern that arises based on the laws of physics and chemistry cannot represent a conceptual abstraction of which it is unaware.

We often hear the term, *emergent properties of the brain*, to account for intelligence and knowledge, but just briefly, what is really meant is emergent properties of the **mind**. You may believe that the mind is nothing more than a description of brain processes but even so, *emergence* requires something from which to emerge, and that something must have properties which are foundational to the properties of that which emerges. Emergence cannot explain its own origins, as we have noted before. To borrow the words of Joseph Keating, emergence "is no more than a 'pseudo-explanation', and may deceive us into believing we have explained some aspect of biology when in fact we have only labelled our ignorance."[57]

Our system of sight is a process by which external light signals are converted to an electro-chemical data stream which is fed to the brain for storage and processing. The data must be encoded in a regulated manner using a protocol that is comprehensible by the recipient. The brain then stores that data in a way that allows correlation and future processing. Evolutionists would have us believe that this highly complex system arose through undirected processes with continual improvement through generations of mutation and selection. However, there is nothing in these processes which can begin to explain how raw data received through a light sensitive organ could be processed without the pre-loading of the meta-data that allows the processor to make sense of the raw data. In short, the only source of data was the very channel that the organism neither recognised nor understood.

Without the meta-data that establishes the relationship between the physical symbols and the abstractions they represent, no meaningful storage or processing can occur. Without the back-end storage, retrieval, and processing of the data, the input device has no useful function. Without an input device, the storage and retrieval mechanisms have no function. Just like a computer system, our sensory sight system is irreducibly complex.

CHAPTER 8

Knowing That You Can

"I think I can," puffed the little locomotive, and put itself in front of the great heavy train. As it went on the little engine kept bravely puffing faster and faster, "I think I can, I think I can, I think I can."
~ The Little Engine That Could, Watty Piper, 1930 ~

Every activity of every organism is underpinned by the "knowledge", conscious or otherwise, that such activity is possible. I cannot be sure whether animals ever attempt activities that may be beyond them: do dogs watch birds and wonder if by launching off a cliff they will achieve flight? Do cats look at a stream and wonder whether breast-stroke may be the most efficient method of swimming? Do birds ever get tired of the cold at altitude and decide to walk south that winter, or maybe just secretly winter over in some human's barn? Do animals even have that remarkable trait of humans: *trying* when failure is a distinct possibility? What particularly bemuses me about those who hold to the evolutionary concept that our free will is but an illusion (see next chapter) nevertheless live as if free will is real and a powerful force for human advancement. Even those scholarly types who advocate the absence of free will are at a loss to explain why they chose the profession they did and why they tried so hard to succeed. If evolutionary processes are indeed absent of purpose or teleological drive, how can this be so?

The issue of self-awareness has long mystified scientists of numerous disciplines, leading to questions regarding the differentiation of the mind and brain. Evolutionists, and I suspect, most neuroscientists, subscribe to the philosophy of *material monism*, the ontological position that all is material and that all phenomena can (and must) be explained in physical or material terms. In that context, "this dominant materialism gives the brain complete superiority over the mind, even in the experience of consciousness"[58]. The opposing position is that the mind controls the brain, or even that the *Self* controls the brain as proposed by John Eccles[59], but that hardly advances our knowledge: what is the nature of the mind - material or non-material? As I have sought to explain in the preceding chapters, an agent external to the material brain is required to manage the multiple levels of abstraction from the symbolic representation of data to cognitive information and knowledge. In other words, the faculties of cognition have yet to be explained in material terms and in my opinion, never can be.

The nature of consciousness, or self-awareness, continues to be the subject of scientific contention, the complexity being such that perhaps it is best left to the experts. In this chapter, however, I would like to consider awareness at a more fundamental level: the interaction of the brain with muscles, tendons, and the like from an evolutionary perspective.

The term "muscle memory", synonymous with motor learning, can be defined as a form of procedural memory that involves consolidating a specific motor task into memory through repetition. All very well, but before a creature can perform such tasks, it must know that it can, and it must have a purpose in mind. As best as I can understand, the brain controls movement of our limbs, apart from involuntary spasms of course. That means that the brain is aware of the limb, its functions, and the possible range of movements. The issue is how to explain the evolution of awareness of an evolving feature such as a fin, arm, leg, wing, or whatever, and the awareness of its potential function.

Stumpy the Fish

Imagine if you will, some primitive fishy creature, more like an eel being bereft of fins, swimming in the waters of the early Earth. A genetic mutation occurs which leads to the growth of a protuberance, either within or upon the external surface of the body. That mutation is subsequently inherited by successive generations and continues to grow until it is quite prominent. By chance, attendant muscles or tendons evolve along with the ever lengthening protuberance even though at that stage no survival advantage is offered. From here, the protuberance could evolve into a fin, a leg, or even evolve no further and just be an ugly lump.

Let us suppose it will become a fin.

At some point Stumpy has to come to the realisation, consciously or otherwise, that this embryonic feature is capable of function: how does that happen? Let us assume that for reasons yet to be uncovered, the muscles or tendons begin to expand and contract. Now either the muscle movement was autonomous and informed the brain, or the muscle moved in response to signals from the brain, although just what the brain intended for the movement is likely unknown to even the brain at that point; after all, it has never had one of these stumps before. Of course, it could have been just an involuntary spasm but that would hardly serve the progress of evolution. One further problem relates to the positioning and symmetry of the stump. If we are speaking of chance mutations, then the stump could have arisen anywhere on the creature, in any number, and if more than one, could have been symmetrical or asymmetrical, or of unequal growth and capability. Perhaps stumps grew all over various progeny until the right combination or positioning favoured selection. As with many other phenomena, evolution must explain symmetry and the optimum placement of appendages.

Evolution would have us believe that for whatever reason, the brain commanded the stump(s) to move in various ways until eventually the movement actually became useful. One could imagine two stumps moving in opposition to one another with Stumpy simply revolving in place, or perhaps the movement was vertical and Stumpy either crashed to the bottom or broached the surface. Perhaps by chance the movement of one or more stumps became coordinated but in any event, what was the mechanism that allowed Stumpy to come to the realisation that such movement was or would become useful? Had Stumpy been quite content to glide gracefully through its watery element for whatever number of generations, should not the addition of protuberances for which there was at that time, no useful function, be seen as degrading rather than enhancing survival fitness? Would that not be like bolting two buckets to the side of an F-16?

Up from the Sea

Evolution would also have us believe that over time, similar protuberances evolved to become legs which enabled fish to leave their natural watery environment and venture onto land in search of whatever. Perhaps as their ponds dried up, something prompted them to follow some form of sensory perception that informed them of the presence of a larger body of water, but all sensory perception needs a method of verification lest what is perceived is at best a mirage or at worst an unrecognised danger. In all likelihood, this could only be accomplished through trial and error, with significantly more errors than successes. This raises the question: what is the probability of survival over the long-term when so many chance occurrences have to work in combination in the right way? There is also the issue of the probability of the embryonic organism that inherited the mutation surviving to the adult stage to be able to continue to develop and pass on the experience.

Quoting from an earlier work on the subject:

> "In offering climatic conditions as a selection pressure that guided evolution, the author [Richard Dawkins] mentions the development of amphibians with the words "Fishes that made their living in water could benefit from a temporary ability to survive on land while they dragged themselves from a shallow lake or pond that was threatened with immanent

desiccation to a deeper one in which they could survive until the next wet season"[60]. At first, reading this seems eminently sensible, and one could not dispute the logic that such temporary ability would indeed provide survival benefit, but if you try to imagine the small, gradual, incremental steps needed to achieve that capability, it becomes a somewhat different problem.

I can only deal with this briefly here, but let me cast some bones that might magically re-arrange themselves in a way that would achieve amphibian evolution. The author acknowledges these issues, in a way, but offers no explanations.

Firstly, we have the fins to legs issue and how to overcome that handicapped stage. Then we have the gills to lungs, or whatever mechanism gets added to extract oxygen from air rather than water, and how that manages to develop: Did it develop to maturity before the exposure to air (i.e., before it offered any selection advantage), or did it continue to develop in parallel with periodic exposure, still with no selection advantage until it actually worked?

How about the skin covering, fish skin/scales offering no ultra violet or other protection from the sun? How about a mechanism for regulating body temperature, not a problem for fish in the same way that it is for land dwellers; eye covering or mechanism to prevent the eyes drying; bowel activity in the absence of water? Maybe none was required.

There is the question of all of these, and perhaps more, mutations happening or accumulating in the same animal at the same time: they may have happened independently in different populations over a period, but for the transition to be successful there has to be a minimum set of capabilities that will allow the fish to successfully navigate dry land. Of course, we could posit that many, if not all, occurred in a semi-wet environment that amphibians first evolved in mud and then gained the capability to tackle the desert, but that alters the problem only in degree.

Next, consider bio feedback mechanisms, the ability of the brain to take information from the organ, understand what it means in its context, process that information, formulate a response, and communicate that information back to either that organ or another with a complementary process, or perhaps both. Genetic mutation of organs and limbs needs to be supported by corresponding mutations in the brain for the animal to know what to do with these new features, and the bio feedback mechanisms need corresponding, simultaneous mutations to offer any survival benefit.

Finally, volition, the mental ability to formulate the thought to crawl from a shallow pond across dry land to another pond: Where did that come from? Putting legs on a fish will not enable it to walk; it has to know that it can walk, it needs to want to walk, and all of this has to be enabled by genetic mutation, presumably in the brain.

Evolution of fish to amphibian is far more complex than evolutionists admit in the generally available literature, and my scepticism does not arise from ignorance or incredulity; it arises from knowledge."[61]

I have since found that the species known as "mudskippers" are even more complex than I knew. They have numerous unique attributes that allow them to survive on exposed mudflats such as stereoscopic vision; liquid-filled skin 'cups' to keep their eyes moist when out of the water; a special retina that allows them to focus through the upper and lower sections of their eyes depending on whether they are in or out of the water; the ability to breathe through their mouth and skin (rather than gills); and a lubricant slime covering with antimicrobial properties[62]. The number of genetic mutations required to achieve that additional functionality must attract serious questions about the ability of a variation-selection mechanism to produce such an animal.

The most problematic issue is this: the evolution of a fin, tail, leg, arm, or whatever had to continue in the right direction for generations with no survival advantage, and before the creature's brain became aware of a possible future function. This is another significant issue which evolution has yet to explain: *complementarity*. When we examine the complexity of an appendage such as a leg, it is valid to ask whether each capability evolved in sequence, and the right sequence at that, or in parallel? The evolutionists posit that each characteristic evolved for a different function, and only later combined, thus contending with the argument of *irreducibility*. That is a fine proposition, and can be validated in some cases, but as an overarching explanation it fails dismally as so many have shown.

If indeed the brain controls the muscles and tendons as modern physiology teaches, what possible mechanism could there have been for the brain to eventually devise the right command sequence to have the feature perform in a useful way, and why would it have done so anyway?

The Incomplete Bird

There is a variation of the albatross called a "gooney bird" which has titillated observers for centuries with its ungainly antics, particularly when attempting to land. As a former military air traffic controller working with ab-initio cadet pilots, I recall how we were simultaneously entertained, amused, horrified, and even frightened at times as the students sought to imitate the much derided gooney bird, albeit unintentionally. Clearly, graceful landing is an acquired skill with humans at least having the knowledge that such is attainable, even if the gooney bird seems not to have such hope.

Evolutionists tell us that birds evolved from some form of dinosaurs or lizards, though some tell it the other way around. No matter, the problem remains the same so let us consider the dinosaur-to-bird hypothesis. The anatomical and physiological differences between dinosaurs and birds have been well documented elsewhere so we have no need to revisit them here, save to note that the differences are significant and the mutations required to achieve the transition must have numbered in the millions at least. If evolution is a gradual process, then the transition to flying must have taken some considerable time. At whatever mutation rate science would posit, this timescale is troubling.

In my book quoted above, *The Dawkins Deficiency*, I coined the term "handicapped fossils" to highlight the duplicity of evolutionist like Richard Dawkins when they assert that there are no "missing links" in the fossil record: that there are sufficient fossils of transitional stages (intermediates) to support the overarching narrative. I called these fossils *handicapped* because there must have been animals where some characteristic was in transition, but was no longer fully functional in its original role, and not yet fully functional in its new role. Given the vagaries of evolution and the extended periods over which transitions would have occurred, handicapped fossils should outnumber fully functional fossils by several orders of magnitude, yet they are entirely absent. Quoting from the book:

> "Whatever dinosaur-to-bird-to–dinosaur evolution scenario one prefers, it must be the case that for an extended period of time in even (claimed) geological timescales, there were lots of intermediates running around that were neither fully functional birds, nor normal functioning dinosaurs.
>
> The evolutionary steps between the two have not been calculated to my knowledge, but it must have involved hundreds of millions of mutations over perhaps thousands of generations, and even these numbers may be understated. It would be entirely optimistic to think that the evolutionary path proceeded in just one direction, that there were no branches in the intermediate stages that led to unsuccessful development, that somehow the genetic development knew where it wanted to go and just went there in incremental steps, one building on the other, each proceeding in the direction that eventually resulted in a bird, or dinosaur, depending on which direction you believe it proceeded.

> It is difficult to imagine that both evolution scenarios occurred; that would require a great deal of explaining. If we accept that evolution proceeds via random mutation and natural selection, there must have been an enormous number of animals––neither fully bird nor fully dinosaur, some leading to evolutionary dead ends, some on the path to success––during a very extended period. So, where are the fossils of these innumerable partly functional animals?
>
> Natural selection would have de-selected lots of evolutionary false starts, and given that intermediates between dinosaurs and birds would have diminished survival capacity, I find it difficult to understand why they were not all de-selected, let alone just some, but apparently they were not. Otherwise, dinosaurs could not have evolved into birds (or birds into dinosaurs)."[63]

Before putting aside those particular difficulties, we should note that in some respects, they are trivial in comparison with the primary issue raised in this chapter: that of knowing that you can, or in the case of proto-birds, knowing that they could fly.

Stumpy must have had considerable difficulty in coming to grips with the potential of his/her evolving protuberances, but at least it was all happening within the one watery environment: not so for the unfortunate birds seeking to leave the safety of Mother Earth. If forelegs evolved into wings, there would have been a long period when the dinosaur's brain still considered them as legs and would have continued to try to use them as before, leading no doubt to great frustration. Attempting to use embryonic wings as legs could have offered no survival advantage whatsoever, quite the opposite in fact, and would argue for deselection. The issue becomes: What caused the dinosaur to give up on leg function and seek another usage for whatever form the appendage that used to be legs now took on? Evolution of a physical appendage is one thing, but along with that process the brain needs to be informed of the new function. The answer is, I suspect: no-one knows. More to the point, nothing in the research I have undertaken gives the slightest hint of how meaning can be imparted to the brain through the claimed evolutionary processes of growing new features such as fins, legs, or wings.

Summary

We could speculate endlessly on the sequence of the evolution of legs to wings, or even the evolution of wings from an entirely different anatomical source, and we could wonder about how long it took for the proto-bird to find the right muscle movement before successfully achieving controlled flight, but it would still be nothing other than speculation. There is a more fundamental problem to be solved: What informed the brain to firstly abandon natural usage and experiment with new muscle movements when it had no idea of what such movements might achieve? If new movements were entirely random, given that evolution is undirected, the likelihood of attaining the right combination of complex movements, by chance alone, is something that only wishful thinking could contemplate.

The phenomenon of *acquired characteristics* remains a subject of contention, largely because it is needed for evolution to be true but the mechanism is (as yet) unknown. It is suspected that epigenetic inheritance through generations may be controlled by long noncoding RNAs, but this is yet to be proven. Curiously, though, Lamarck may have been onto something even though he has long been refuted. Even if true, the difficulty for the evolution narrative is that undesirable traits have the same likelihood of inheritance as desirable traits until deselected in later generations. Given that in the case of the proto-bird with their unintentional actions, unsuccessful experiments with proto-wing movement would far outnumber successful outcomes. In my mind's eye, I see generations of clumsy proto-birds trying to get airborne but crashing back to earth like the unfortunate gooney bird without achieving much at all. The Wright Brothers first flight was monumental by comparison!

Humour aside, evolution proponents must come to grips with this important issue if their narrative is to be in any way plausible. As best we know today, the brain controls muscle movement and for it to do that effectively, it must "know" that it can and must "know" that such movement will be functionally useful. We also have learned from amputees that the brain retains such "knowledge" long after it has been proven to be false. If fins, legs, arms, and wings did indeed evolve, there would need to have been a process or mechanism to inform the brain of their existence, and an impetus within the brain to cause their movement. According to the evolution narrative, there is no ontological basis for such activity - it is all simply a matter of chemical interactions, which raises the question as to what chemical reactions would arise in an undirected and autonomous manner to cause muscles to move in specific ways.

Finally, if muscle memory is a function of DNA, the multiplicity of possible evolutionary pathways from arms, or from nothing, to fully functioning wings, arms, or legs might well have left a legacy of false starts in the form of truly junk DNA including that which controls the brain. Should there be traces of "muscle memory", the legacy of repetitively performing certain actions even where such actions proved entirely useful or futile. It would be implausible to suggest that all traces of such false starts disappeared with the less survivable forms of the creatures, and only the successful have survived but somehow absent of any trace of the failures along the way. Current research is not finding any such evidence, or perhaps it is not even looking.

CHAPTER 9

Volition and Free Will

"We must believe in free will, we have no choice"
~ Isaac Bashevis Singer ~

Anthony Cashmore, a Biology Professor at the University of Pennsylvania, assures us that consciousness and free will are simply illusions. He wrote: "The reality is, not only do we have no more free will than a fly or a bacterium, in actuality we have no more free will than a bowl of sugar."[64] Charles Darwin, Thomas Huxley, Francis Crick, and numerous other evolutionists are of the same view, Susan Blackmore observing, "I think nature has played this enormous joke on us."[65] Some scientists and philosophers have taken this a step further, positing that as we do not really have free will, we are thus not responsible for our own actions: our behaviour is simply the result of complex chemical reactions over which we have no control. William Provine, a Biology Professor at Cornell University commented: "There is no way that the evolutionary process ... can produce a being that is truly free to make choices."[66]

Taking Professor Provine at his word, if it can be demonstrated that we <u>can</u> make free choices, as he apparently did when making his assertion, would he then agree that evolution theory must be false?

Does Professor Provine truly believe that his choice of profession, his dedication to hours of study and research, and the conclusions that he has reached, are all the results of complex chemical reactions over which he has no control? Should any of these academics be lauded for their achievements when according to them, they had no personal influence over the outcomes? Should the Nobel Prize ever be awarded to anyone when they are little better than chemical automatons?

As these scientists are talking about themselves as well as everybody else, we cannot ignore the obvious: their own theory strips evolutionists of any authority to make such assertions. One would hope that such scientific luminaries are joking when they deny the existence of volition, but I suspect not.

Gilbert Ryle, former Waynflete Professor of Metaphysical Philosophy at the University of Oxford noted way back in 1949:

> "Teachers and examiners, magistrates and critics, historians and novelists, confessors and non-commissioned officers, employers, employees and partners, parents, lovers, friends and enemies, all know well enough how to settle their daily questions about the qualities of character and intellect of the individual with whom they have to do. They can appraise his performances, assess his progress, understand his words and actions, discern his motives and see his jokes. If they go wrong, they know how to correct their mistakes. More, they can deliberately influence the minds of those with whom they deal by criticism, example, teaching, punishment, bribery, mockery and persuasion, and then modify their treatments in the light of results produced.
>
> Both in describing the minds of others and in prescribing for them, they are wielding with greater or less efficiency concepts of mental powers and operations. They have learned how to apply in concrete situations such mental-conduct epithets as 'careful', 'stupid', 'logical', 'unobservant', 'ingenious', 'vain', 'methodical', 'credulous', 'witty', 'self-controlled' and a thousand others."[67]

If you consider these observations in the light of humans having no free will, you should be able to see that there is something seriously wrong with the perspective of the evolutionists. Could any of these phenomena occur in the absence of free will?

There is, perhaps, no empirical evidence of volition and free will, but there is certainly an abundance of circumstantial evidence within the lives of every one of us, and of every person recorded in human history. What physical properties, arrangements, or processes of the human brain result in choices, decisions, preferences, fancies, or predilections? Consider twin boys: from an early age, one wants to study history and dreams of tenure in academia, whilst the other sees himself as a fighter pilot and a lifetime as an aviator; on the basis of an almost identical genome, what physical properties are responsible for the differences and almost opposite polarity in the choice of careers? When aspiring pilots failed to demonstrate adequate proficiency, as I and many others did, why did some continue to pursue careers in aviation whilst others chose accountancy, engineering, or alternate careers not at all connected to aviation? Raised as a Catholic, I pursued independent bible studies and eventually came to the conclusion that I no longer accepted all of the doctrines of traditional Christianity, yet many of my peers remained as devout Catholics, whilst others experimented with Buddhism and other belief systems. What was it in the physical make-up of our brains that predetermined these separate outcomes?

Why is it that some individuals from generations committed to one religion eventually chose to abandon that religion, and even their children and families, to follow another path as Carma Naylor[68], Tony Coffey[69], Jerry Rassamni[70], and Walid Shoebat[71] did? Why did some atheists choose to accept the existence of God contrary to their previous teachings and beliefs as Francis Collins[72], Anthony Flew[73], Peter Hitchens[74], C.S. Lewis[75,76], John Sanford[77], Lee Strobel[78], and many others have publicly attested? Why did an Egyptian woman whose father was assassinated by the Israelis eventually establish an organisation supporting Israel contrary to her culture and her tragic experience at the hands of the Israelis[79]? Were these people, and the multitudes of whom we have never heard, all compelled by complex chemical reactions to choose as they did?

Did Tchaikovsky compose the *1812 Overture*, did Rembrandt paint *The Night Watch*, did Michelangelo sculpt *David*, or were these famous artists unwitting accomplices to chemical reactions over which they had no control?

The very same people who are proponents of evolution, also as parents and teachers exhort us to develop character? How can we develop character if we have no control over who we are and how we react to circumstance? Why are some people honest and others dishonest; some gentle and others brutal; some altruistic and others selfish? Why do courts and secular authorities believe that criminals can be rehabilitated if the criminals have no control over their future behaviour? Why did despots such as Lenin, Trotsky, Stalin, Hitler, Himmler, Pol Pot, and Mao Zedong all believe that indoctrination was an effective tool in bending people to their will? Why do commercial enterprises spend exorbitant funds trying to convince people to buy their products if nobody truly has free will and the ability to choose? Why do universities have courses in political science, philosophy, psychology, psychiatry, psychotherapy, ethics, and related subjects if the imparted knowledge is incapable of changing behaviours?

How do people live their lives - do their actions testify to the assertions of the evolutionists, or to some other belief system? How do the evolutionists live their lives - do they live as if they truly believe what they teach, or do they give the lie to their beliefs and live as intellectual hypocrites? Why would anyone attend university or pursue a career when if the evolutionists are right, they can have no confidence that the chance chemical reactions in their brains won't have them heading off in some other direction just moments later? Why would anyone take out a loan and invest tens of thousands of dollars in an education when they can have no assurance that their dreams won't be shattered by a later chemical reaction? What is the point of following a hope, dream, or ambition if indeed you have no personal relationship to it other than an evolutionary event caused it to intervene in your life?

Summary

I have no doubt that chemical interactions are involved in the implementation of decisions - it must be so otherwise I could not be typing the words that I choose to type. I also accept that cognitive functions are in some way related to electro-chemical reactions in the brain, but the question remains, what is driving what, or as Lewis Carroll expressed in another context: '*The question is,*' said Humpty Dumpty, '*which is to be master — that's all.*" Have I deliberately chosen to write what I have written, or has the choice been made for me and I only write because I am compelled to write, what I write, when I write, due to complex chemical reactions over which I have no control?

Are all our actions, words, decisions, choices, and preferences merely the outcome of complex chemical reactions over which we have no control, or do we in truth have some level of control expressed as *free will* or *volition*? If we have no free will, should anyone be held accountable for anything: why should achievers be rewarded and offenders punished?

We are faced with a *contradictory reality*. On the one hand the evolutionists, not being able to explain free will and volition in evolutionary terms, have chosen to deny that such exist even as they practice those attributes when they profess their denials. On the other hand, we have ample evidence that free will and volition do exist even though we are unable to explain them in physical terms. It seems prudent, to my mind, to accept the existence of free will and volition and to ponder whether their nature is immaterial. If that be the case, the logical corollary would be that we have an immaterial mind that controls the brain in all but autonomous functions.

The denial of humans having free will is so absurd and so contrary to the evidence that at best we should describe this assertion by evolutionists as an *argument from desperation*. In any event, if we have no control over our mental processes, then clearly evolutionists can have no confidence in what they believe or preach, and nor can we.

Revisiting the observation by William Provine that "There is no way that the evolutionary process … can produce a being that is truly free to make choices", I find myself free to make the choice that the evolution narrative must be false.

CHAPTER 10

Distinguishing Mind from Brain

"Bodies devoid of mind are as statues in a market place"
~ Euripides ~

As noted earlier, evolutionists will generally assert that "the mind is an emergent property of the brain", whilst in truth having no science to explain this phenomenon. One observation that can be offered, however, is that as many scientists do distinguish the *mind* from the *brain*, it is thus incumbent upon the scientific community to identify which properties or characteristics are separately discernible. I believe it to be useful here to invoke Leibniz's law of the *indiscernibility of identicals*. J.P. Moreland puts it this way:

> *"In general, if 'two' things are identical, then whatever is true of the one is true of the other, since in reality only one thing is being discussed. However, if something is true of the one which is not true of the other, then they are two things and not one. This is sometimes called the indiscernibility of identicals and is expressed as follows:*
>
> $$(x)(y)[(x=y) \rightarrow (P)(Px \leftrightarrow Py)]$$

For any entities x and y, if x and y are really the same thing, then for any property P, P is true of x if and only if P is true of y."[80]

To apply these principles to the mind versus brain conundrum, we need to approach the problem from a different perspective, lest we simply affirm the consequent. Rather than start with the presupposition of separateness, we need to identify specific properties and ask of each: can this be a property of a physical organ such as we know the brain to be, or are these characteristics of a phenomenon incapable of explanation by the physical sciences? If all properties or phenomena can be explained by reference to the material brain itself, then no discernible separateness can be asserted.

It is not my intention here to review an exhaustive list of what we know about the brain and related phenomena: such is beyond my knowledge and I would leave such investigations to those better qualified. However, we can inspect some of the more commonly recognised phenomena to demonstrate the process of determining whether or not the mind and the brain are separate entities. The table at the end of this chapter illustrates my thinking, and I will leave it to you, the reader, to agree or disagree with the categorisations, but before we get to that, I would like to introduce some of the thinking of recognised experts in the field to discover whether they address any of my core issues.

Karl Popper & John Eccles

I was keen to study the 600-page volume, *The Self and its Brain*[81], by these authors but very quickly my enthusiasm turned to disappointment: the authors presupposed the truth of evolution and sought to describe (rather than explain) the subject in that context. Animal sentience, human consciousness of self and death, human language, works of art, all "events of creative evolution or of emergent evolution"[82]. *Emergent*: there is that word again, as if there was some scientific principle behind it. In fairness, the authors did acknowledge the objections of those who argue that "if the universe consists of atoms or elementary particles, so that all things are structures of such particles, then every event in the universe ought to be explicable, and in principle predictable, in terms of *particle structure* and of *particle interaction*."[83] [italics in original] We should hold onto that thought, for if all of reality is contingent upon particles and their structures, or as some have put it, everything is material, then evolutionists must identify those particles and their structures, and demonstrate not just the interactions within the brain that give rise to such behaviour as volition and creativity, but also the causality.

Exploring the views of others, the authors noted: "some of the most important of living philosophers (such as Quine) teach that there can be only physical entities, and that there are no mental events or mental experiences. (Some others compromise and admit that there are mental experiences, but say that these are, in some sense, physical events, or that they are 'identical' with physical events.)"[84] As I sit here typing these words, I am compelled to ask: if not mental events, what are these events that I am experiencing, including this event where I ask myself the question? How would Quine explain his reasoning if not in terms of mental events?

There is an intriguing section in this referenced book on brain-mind interaction, but as the authors admit, it is an hypothesis. There can be no doubting that there is, and must be, an interaction between the mind and the brain, but as my research has revealed, there is little if any consensus on the mechanisms. The neurosciences have been able to show where certain activities occur in the brain, but have been unable to identify the causal factor other than when external stimuli are involved. For example, brain activity can be identified when a person moves an arm or a leg, but there is no physical evidence of the *act of deciding* to move. In other words, there is no physical evidence of the volition that precedes the act, casting doubt on the notion that volition is a physical event.

It would be both unfair and irresponsible of me to condense the work of these authors into just a few examples favourable to my argument, but as nowhere in their study did they address the fundamental difference between the physical and the conceptual, there is no point in reviewing their work further. In the context of my study, we have already identified an important point: the authors assume evolution and seek to explain the mind in terms of the physical properties of the brain. Despite an impressive body of research and an erudite treatment of the subject, fully supported by the works of qualified scientists and others, they add nothing to my understanding because of their foundational assumptions. In respect of the *mind*, the conclusions drawn and explanations offered are in truth assumptions, despite the verifiable truths of some aspects of the neural sciences.

I do not belittle the works of the neuroscientists who have made such wonderful progress in understanding the functions of the brain, and in identifying the areas of the brain where neural processes are occurring. But the question remains: are they seeing the *cause*, or the *effect*?

The physical sciences rely on the principle of reductionism: everything that happens at one level can be explained by causal events at a lower level. In the context of my study then, composing a poem or song can supposedly be explained by neural interactions in the brain in turn caused by chemical and physical interactions caused by ...... what? What is the trigger event or sequence of events that resulted in a poem instead of a song?

Gilbert Ryle

My reference text for this author is *The Concept of Mind* quoted in earlier chapters. Ryle's approach is thought provoking, for he contends:

> "Descartes left as one of his main philosophical legacies a myth which continues to distort the continental geography of the subject. A myth is, of course, not a fairy story. It is the presentation of facts belonging to one category in the idioms appropriate to another. To explode a myth is accordingly not to deny the facts but to re-allocate them. And this is what I am trying to do."[85]

Ryle goes on to note that there has been, at least since the time of Descartes, what he describes as the *official doctrine*. In essence, this doctrine is that "human bodies are in space and are subject to the mechanical laws which govern all other bodies in space ... but minds are not in space nor are their operations subject to mechanical laws."[86] However, according to Ryle, this doctrine is absurd. He states:

> "I hope to prove that it is entirely false, and false not in details but in principle. It is not merely an assemblage of particular mistakes. It is one big mistake and a mistake of a special kind. It is, namely, a category-mistake. It represents the facts of mental life as if they belonged to one logical type or category (or range of types or categories), when they actually belong to another. The dogma is therefore a philosopher's myth."[87]

Encouraged by this logical approach, and unsure of where he was heading with his argument, I continued to read through the text, looking for justification of why the dogma had stumbled over a category-mistake, and in which category our mental life truly belonged. Ryle presents some wonderful chapters on the will, emotion, self-knowledge, imagination, and the intellect, all of which are deserving of deeper study, but not once does he address the central questions that had me select his book in the first place: are the mind and brain separately discernible with different properties, or are they essentially the same, and if they are the same, how are the "mental" and conceptual aspects explained in physical terms? I am very glad that Ryle wrote this book, and I intend to return to it at a later date on a different quest: his chapter on "knowing that" versus "knowing how" is particularly insightful, but for now I have put it aside as not relevant in this context.

But back to categories for a moment. Categories can be useful, but rather like statistics, they are often used as a drunk uses a lamp post: more for support than for illumination. As a former data analyst and data base designer, I often faced the same problem as librarians: how do you best categorise an entity when its subject spans multiple established categories? In my home budget, does my vehicle insurance go under "insurances" or under "vehicle costs"? When I travel by train specifically for a medical appointment, does the cost go under "miscellaneous transport" or under "medical costs"? In brief, categories are generally defined based on their intended usage, so when someone asserts a *category mistake*, one should ask: on what basis were the categories defined? Categories are based on a particular perspective – a different perspective will usually result in different categorisations, and so it is in this instance. Notwithstanding Ryle's insights in many areas of intellect and cognitive processing, a lack of definition regarding false categorisation leaves his arguments on a questionable foundation.

Returning just briefly to the "official doctrine", by *mechanical laws* we now understand genetics, microbiology, biology, physiology, and so forth - all of those applicable scientific disciplines which nevertheless all fall within the category of the physical sciences. I would offer that despite the uncertainty of exactly where (or when), every physical entity, no matter how small, must be somewhere. In the case of thoughts, given that they are not externally discernible, they must, if they are physical, reside somewhere in the brain, as a construct or assemblage of molecules, or as a pattern of connections. It could just be early days in the science, but insofar as I have been able to discover, there is no understanding of what any particular thought "is" even though it can be seen to be occurring. What can be seen in the brain is a coded pattern of activity, which leaves scientists the task of discovering their own version of the Rosetta stone.

Similarly, information and knowledge, if they are physical, must reside in the brain, but as I have earlier argued, information and knowledge are *conceptual*, not *physical*, and are thus not constrained by a physical location or structure of locations.

Despite opening with his rebuttal of Descartes' myth, Ryle never returned to this issue of the mind not being in space. In fairness though, if any criticism is to be offered, it relates more to my expectations than to the author's writings.

Daniel C. Dennett

My reference text for this author is *Consciousness Explained*[88]. Again, my interest was not so much to understand consciousness in all its complicated detail, but to seek arguments for or against the mind and brain being one. Can the mind be an emergent property of the brain, or are the two intrinsically different? Can physical evolution truly account for a realisation of the conceptual, and if so what are the physical mechanisms?

Dennett early states:

> "The idea of mind as distinct ... from the brain, composed not of ordinary matter but of some other, special kind of stuff, is dualism, and is deservedly in disrepute today ... the prevailing wisdom, variously and expressed and argued for, is materialism: there is only one sort of stuff, namely matter – the physical stuff of physics, chemistry, and physiology – and the mind is somehow nothing but a physical phenomenon; in short the mind is the brain. According to the materialists, we can (in principle) account for every mental phenomenon using the same physical principles, laws, and raw materials that suffice to explain radioactivity, continental drift, photosynthesis, reproduction, nutrition, and growth. It is one of the main burdens of this book to explain consciousness without ever giving in to the siren song of dualism. What then, is so wrong with dualism? Why is it in such disfavour?"[89]

At last, I thought to myself, here is an author willing to tackle this issue head-on. He raises the very real problem (as did the other authors incidentally) of how a non-material entity could have influence over a material one. He notes:

> "A fundamental principle of physics is that any change in the trajectory of any physical entity is an acceleration requiring the expenditure of energy, and where is this energy to come from? It is the principle of the conservation of energy that

accounts for the physical impossibility of 'perpetual motion machines', and the same principle is apparently violated by dualism. This confrontation ... is widely regarded as the inescapable and fatal flaw of dualism."[90]

Within the paradigm of materialism, I could not agree more, but there is a flaw in the argument, one very similar to that used by Richard Dawkins when he asks: "Who made God?" The proponents of dualism contend that the mind is non-material: thus to use arguments based on physics, which deals only with the material, is to beg the question. If the non-material does in fact exist, the relationship between it and the material cannot be explained in purely material terms - that much should be obvious. Physics, by definition, can never explain how a non-material entity could have influence over a material one – some other category of reality must be investigated.

The remainder of Dennett's 500-page volume is predicated on the presupposition that everything is material, and thus every explanation must be sought within that paradigm. His ideas are very interesting and in parts imaginative, but he does not address what to me are core issues relating to the conceptual nature of information and knowledge. In many ways, his arguments suffer from the same weakness as described by Professor Andrews regarding the wilful blindness of atheists:

> "A further example of circular argument is the idea promoted by some atheists that 'science disproves the existence of God'. The assertion is based on the claim that science presents no evidence for the existence of supernatural forces or phenomena. It sounds plausible until you look a little more closely. The argument can be expressed as a syllogism as follows:
>
> 1. Science is the study of the physical universe.
> 2. Science produces no evidence for the existence of non-physical entities.
> 3. Therefore, non-physical entities such as God do not exist.

Again the fallacy is clear. In point (1) 'science' is defined as the study of the physical or material world. This statement thereby excludes by definition any consideration by science of non-physical causes or events. The proposition then argues from the silence of science concerning non-material realities that such realities do not exist. By the same logic, if you define birds as 'feathered creatures that fly', there's no such thing as an ostrich. It's fairly obvious in this example whose head is in the sand."[91]

Summary

I trust that I have not been unfair, nor have done a disservice to the authors reviewed above, but as much as their works have been informative, none to my mind have investigated the nature of information and knowledge as I have attempted to do in this book. It is one thing to try to explain these concepts in materialistic terms, but without defining what it is that one is dealing with, such explanations must always be questionable. If information can be manipulated by physical activities in the brain, these authors should have shown the relationship between the physical symbols or expressions and the data that such expressions represent. It is all very well to talk of correlations, but correlations of what exactly: electrical gradients? The only aspect of brain activities that are truly empirical are physical in nature, but missing is the intelligent coding structure that allows physical representations to be expressed at a conceptual level.

I have been struggling to find the right words, or analogies, to illustrate how the mind and brain are separately discernible due to the properties that they exhibit and the entities with which they deal. But let me contend that just as you can physically examine a music CD but never find music there, you can examine the physical brain and never find the conceptual there: one can never find the thoughts, ideas, memories, moral values, religious beliefs, music preferences, or other cognitive knowledge that we all know that we have.

An exercise for the reader

Based on my arguments in the earlier chapters, here are some suggested talking points. If you are familiar with the works of authors like those reviewed above, ask yourself whether they have satisfactorily explained why all of these phenomena can be explained by physical processes within the brain alone.

Property or Phenomenon	Brain	Mind
Feeling of pain or heat	X	
Control over reaction to pain or heat (flinch, cry out)		X
Running, walking, jumping, hopping	X	
Decision to run or walk, skip or hop		X
The act of swimming	X	
Overcoming a fear of water		X
Decision to swim breaststroke, backstroke, etc.		X
Feeling a toothache	X	
Decision to visit dentist, which one, and when		X
Decision to take out medical/dental insurance		X
Become pregnant	X	
Decision to abort or carry a baby to full term		X
Choice of school subjects, college course, and career		X
Get sunburnt	X	
Decide to cover up, apply sunscreen, or leave		X
Choice of which book to read, or not read		X
Choice of which movie or TV program to view		X
Decision to volunteer or give to charity		X
Author a poem, essay, book, song, lyric, music		X
Design a dress, car, painting, sculpture		X
Decision to read this book		X

Evolutionists argue that there are no such individual phenomena as volition, free will, or choice, and consequently no individual can take pride in individual accomplishments of any sort. No individual chooses to work harder, try harder, run faster, jump higher, achieve higher academic qualifications, or in any way be competitive as a matter of personal preference, nor can any individual choose to override a natural instinct or overcome fear or circumstance.

Acknowledging that such ideas may be considered offensive, particularly by high achievers in practically any field, we should nevertheless strive to remove emotion from our deliberations and view the issues objectively. Of course, if you are able to do that, it would suggest that the evolutionists are wrong all along.

A Final Thought

C.S. Lewis, a much respected author, made the following comment on the process of thinking. Although it was in an entirely other context, the logic still holds:

> "Supposing there was no intelligence behind the universe, no creative mind. In that case nobody designed my brain for the purpose of thinking. It is merely that when atoms inside my skull happen for physical or chemical reasons to arrange themselves in a certain way, this gives me, as a by-product, the sensation I call thought. But if so, how can I trust my own thinking to be true? It's like upsetting a milk-jug and hoping that the way the splash arranges itself will give you a map of London. But if I can't trust my own thinking …."[92]

CHAPTER 11

Does the Brain Store Memories?

Wanting to provide the perspective of a medical specialist practising in the neurosciences, I have sourced the following article from here:

http://www.evolutionnews.org/2014/12/recalling_nanas091821.html

The author, Michael Egnor, is a Professor of Neurosurgery at Stony Brook University Hospital, and in the interests of full disclosure, he is also an intelligent design supporter who writes for the Discovery Institute blog. I am delighted that he has so graciously consented to my reproducing his article in full here, and wish to thank him for his contribution.

Recalling Nana's Face: Does Your Brain Store Memories?

"A singular consequence of the materialist-mechanical metaphysics that permeates our culture and our sciences is that we commonly hold basic beliefs that are abject nonsense. One such belief is the almost ubiquitous one -- among ordinary folks as well as neuroscientists and surprisingly many philosophers -- that the brain "stores" memories. The fact is that the brain doesn't store memories, and *can't* store memories.

It has been known for the better part of a century that certain structures in the brain are associated with memory. The *amygdala* and the *hippocampus* in the temporal lobe, and some adjacent cortical regions, have been shown to be associated with the act of remembering in animals and humans. The research is fascinating and important, and in my own work as a neurosurgeon I have to be aware of these regions (especially the hippocampus and the fornix and mammillary bodies, to which the hippocampus projects). During surgery, injury to these critical structures (if bilateral) can leave a patient incapable of forming new memories, which is a crippling disability.

But these physiological facts do *not* imply that the brain stores memories in the hippocampus or amygdala or elsewhere. How so?

It's helpful to begin by considering what memory is -- memory is retained knowledge. Knowledge is the set of true propositions. Note that neither memory nor knowledge nor propositions are inherently physical. They are psychological entities, not physical things. Certainly memories aren't little packets of protein or lipid stuffed into a handy gyrus, ready for retrieval when needed for the math quiz.

The brain is a physical thing. A memory is a psychological thing. A psychological thing obviously can't be "stored" in the same way a physical thing can. It's not clear how the term "store" could even apply to a psychological thing.

Now you may believe -- as most neuroscientists and too many philosophers (who should know better) mistakenly believe -- that although of course memories aren't "stored" in brain tissue per se, *engrams* [Ed: a *hypothetical* means by which memory traces are stored] of memories are stored in the brain, and are retrieved when we remember the knowledge encoded in the engram. Indeed, neuroscientists believe that they have found things in the brain very much like engrams of some sort, that encode a memory like a code encodes a message.

But that too is nonsense. To see why, consider a hypothetical "engram" of your grandmother's lovely face that "codes" for your memory of her appearance. Imagine that the memory engram is safely tucked into a corner of your superior temporal gyrus, and you desire to remember Nana's face. As noted above, your memory itself obviously is not in the gyrus or in the engram. It doesn't even make any sense to say a memory is stored in a lump of brain. But, you say, that's just a silly little misunderstanding. What you really mean to say is that the memory is *encoded* there, and it must be accessed and retrieved, and it is in that sense that the memory is stored. It is the engram, you say, not the memory itself, that is stored.

But there is a real problem with that view. As you try to remember Nana's face, you must then locate the engram of the memory, which of course requires that you (unconsciously) must remember where in your brain Nana's face engram is stored -- was it the superior temporal gyrus or the middle temporal gyrus? Was it the left temporal lobe or the right temporal lobe? So this retrieval of the Nana memory via the engram requires *another* memory (call it the "Nana engram location memory"), which must itself be encoded somewhere in your brain. To access the memory for the location of the engram of Nana, you must access a memory for the engram for the location for the engram of Nana. And obviously you must first remember the location of the Nana engram location memory, which presupposes another engram whose location must be remembered. Ad infinitum.

Now imagine that by some miracle (materialist metaphysics always demands miracles) you are able to surmount infinite regress and locate the engram for Nana's face in your superior temporal gyrus (like finding your keys by serendipity!). Whew! But don't deceive yourself -- this doesn't solve your problem in the least. Because now you have to decode the engram itself. The engram would undoubtedly take the form of brain tissue -- a particular array of proteins, or dendrites or axons, or an electrochemical gradient of some specific sort -- that would mean "memory of Nana's face." But how can an electrochemical gradient represent a face? Certainly an electrochemical gradient doesn't *look* like grandma -- and even if it did, you'd have to have a little tiny eye in your brain to see it to recognize that it looked like grandma. Whatever form the engram takes must be a code, and you must then have a key to the code, stored in your brain just like the Nana memory is stored. But then you must remember where the key to the code is stored, which is itself another memory which must be stored and remembered. And to remember the location of a location for the key for the code for the engram requires another engram to remember the location of the location code, which must be located and decoded, which requires another key engram which you now must locate...

And if you think that remembering your grandmother's face via an engram in your brain entails infinite regress, consider the conundrum of remembering a *concept*, rather than a face. How, pray tell, can the concept of your grandma's justice or her mercy or her cynicism be encoded in an engram? The quality of mercy is not strained, nor can it be encoded. How many dendrites and axons for mercy?

You see the nonsense.

To assert that memories are stored in the brain is gibberish. And don't fall for the materialist invocation of promissory materialism -- "It's just a limitation of our current scientific knowledge, and we promise that science will solve the problem in due time." The assertion that the brain stores memories is *logical* nonsense that doesn't even rise to the level of empirical testability.

How then, you reasonably ask, can we explain the obvious dependence of memory on brain structure and function? While it is obvious that the memories aren't stored, it does seem that some parts of the brain are necessary ordinarily for memory. And that's certainly true. But necessary does not mean sufficient. There is a rough correspondence between activity in certain regions of the brain and the exercise of certain mental powers. That is what cognitive neuroscientists properly study. In some cases, the correspondence between brain and memory is one of tight necessity -- the brain must have a specific activity for memory to be exercised. But the brain activity is not the *same thing* as the memory nor does it make any sense at all to say the brain activity *codes* for the memory or that the brain *stores* the memory.

What this all implies is that only some kind of dualism can provide a coherent understanding of the mind. But dualism is a many-headed hydra, and I don't think that Cartesian dualism or property dualism or epiphenomenalism or computational theories of the mind (which are inherently dualistic) explain things well either.

I hew to Thomistic dualism, which is a coherent view of the mind that takes an Aristotelian perspective and for which the participation of the brain in memory is not problematic at all."

CHAPTER 12

Speaking, Writing & Reading

It is indisputable that cognition is a prerequisite for being able to speak, write, and read, and I place these faculties in that sequence quite deliberately. That said, in an evolutionary pathway, it might appear as not necessarily so. Without speech, an early man or woman may see an animal and attempt to reproduce it in a drawing, thereby communicating the event to another person. In that sense, drawing is equivalent to writing, and observing a drawing meant to communicate is equivalent to reading. But this process can only take communication so far. If I drew an elephant or hippopotamus, I doubt that anyone would recognise either drawing. The point is that the drawer (writer) must perform the task with a certain level of skill, and that the observer (reader) must have previously seen what the drawing is meant to represent otherwise no intelligent communication can occur.

To bridge the gap between the writer and reader, speech is usually employed, and this is the process we experience in learning. Except at a very primitive level of drawings, I believe it to be axiomatic that no-one can learn from any writing without the verbal input of another intelligent entity.

Speaking of Speech

Researching this issue, I have found general agreement that whilst ancient civilisations had spoken language without writing, none have been found where they had the written but not the spoken. How scientists have determined this is beyond my ken, but I shall accept it as true. According to this early guide for teachers[93], learning to speak must precede learning to read. It is from speaking that we first learn vocabulary and speech patterns, and thus acquire the ability to understand what we are being taught orally. Evolutionists have no answer for how speech arose, and according to anthropologist, Alfred Kroeber, how language came about is "one of the darkest areas in the field of human knowledge"[94].

Though I cannot quote the source, I recall reading that if children have not heard a human voice by their early teens, they are unlikely to ever learn to talk.

Writing on Writing

Researching this issue, I have found that scientists in the appropriate disciplines similarly know little about how the art/discipline of writing came about[95]. If we examine the development of an ancient language such as Hebrew, we find that pictograms were initially used because to an extent, such could be comprehended without explanation. An inherent limitation was that the available characters were all uniquely significant, but the number of concepts to be conveyed far outnumbered the comprehensible combinations. Over time, the "alphabet" morphed from pictograms to non-significant characters opening the way for a far wider vocabulary and consequently, the ability to express a far wider range of concepts. In most languages, a large percentage of words have a semantic range, i.e. more than one meaning. In some cases, the meanings are entirely different, and in other cases, they express a nuance or subtlety that another word does not convey. Again I would stress that such variations express concepts, which themselves can only be intelligently derived.

You might cast your mind back to an earlier discussion on poly-functional and poly-dependent. Written language evolved by moving from pictograms (single function) to non-significant characters (poly-functional). As the written symbols became poly-functional, they also became poly-dependent in the sense that additional mechanisms (code books, dictionaries) were necessary to explain what any combination of symbols meant. Pictograms were, in a sense, physical, because they could be easily mapped to an existing reality, but arbitrary symbols are conceptual, requiring an intelligent agent for their implementation and usage.

This is an important issue to understand, particularly as it relates to the subject of origins. As with speech, writing involves complex, multi-level, coding systems devised to express a wide range of concepts. These coding systems have evolved over time, not through undirected organic processes, but through the directed application of cognitive processes intent on solving particular problems or achieving specified goals. It has taken time to learn, teach, and develop further, and even today, new words become necessary to express new concepts, or to express existing concepts in different ways.

In the development of any coding system, there is a direct relationship between the significance of the individual symbols and the number of concepts which can be expressed. Similarly, there is a direct relationship between the number of symbols employed in the coding system, and the number of different messages which can be conveyed. In mathematics, each symbol such as $\sqrt{}$, $+$, or $\div$ is uniquely significant in that it can have just one meaning, otherwise computations would fail. In the binary notation, the symbols 1 and 0 each have just one meaning, but the more of them employed in a string, the more concepts can be expressed; note however that such is the result of a higher level coding system. In a written language such as English, code strings (words) are comprised of letters of the alphabet which as used today, are non-significant symbols (with a few exceptions). These code strings are extended by the use of spaces and punctuation, allowing further concepts such as phrases, sentences, paragraphs, and so forth. Whenever we encounter a concept, we should remember that such can only have their origin in intelligence, not chemicals.

This is how language has developed, both spoken and written, but it has been the cognitive abilities of humans which has allowed this development. In other words, intelligence was the prerequisite for language. Again, from the perspective of origins, how could undirected organic evolution account for such complex intelligence?

As the written language was developed, communication of the changes most likely would have been oral, for no reason other than it seems to work best that way, and new writings must always precede new readings, hence my earlier sequencing of these faculties.

Reading about Reading

Reading is an incredibly complex yet entrancing process, an elegant performance composed by an unknown choreographer bringing together the eye, cognition, memory, and emotions. Not only can we read, but we can even learn to *speed read* which whilst highly lauded in some circles, leaves me feeling that much of the beauty of prose is lost. The process is such that Albert Einstein is said to have described it as "the most complex task that man has ever devised for himself"[96]

As we have noted in other contexts, there are issues of symbols, coding systems, concepts, abstractions, correlations, generalisations, vocabulary, ignoring misspellings, and other cognitive processes which are all employed simultaneously at amazing speeds. Each of these issues comes with its own levels of complexity. For example, as noted in digital information processing, correlation of concepts requires a systematic organisation of data supported by a method of indexing that allows instant retrieval. The self-organisation of chemicals cannot occur in such a manner as to express concepts of which it is not, and cannot, be aware. Thus an external intelligent agent is required to orchestrate and harmonise the contribution of the many faculties which enable reading.

Summary

This has been but a brief overview of what you probably already knew, but a review allows us to consider these issues in the context of origins, and thus evolution. Everything that we know about speaking, writing, and reading is in the context of cognition, so to investigate the origin of any of these faculties requires us to investigate the origin of human cognitive processes. Can undirected, organic, evolutionary processes account for cognitive processes, and if not, they cannot account for our ability to talk, read, or write. Some may consider this nothing but circumstantial evidence against the case for evolution, but I believe it to be very strong evidence.

As George Orwell noted: "We have now sunk to a depth at which restatement of the obvious is the first duty of intelligent men." In the context of circumstantial evidence, Thoreau wrote: "Some circumstantial evidence Is very strong, as when you find a trout in the milk."

We have found not just trout, but an entire aquarium in the milk of the evolutionist.

CHAPTER 13

Understanding "Self"

If all is material and all actions and decisions are simply the result of uncommanded, complex chemical reactions, then effectively there is no such entity as a personal "I" or "You". In essence, we are simply organisms, perhaps more complex than bacteria, worms, jelly fish, and the like, but of no more significance. We are all different, but we play no part in our differences – they are the result of the hand we are dealt when we were born, simply a function of our parents' genes and circumstances. If that is not your personal experience of self, then perhaps you should rethink the assertions of the material-monists and evolutionists, and how much truth there is in them.

I will not dwell on this issue apart from one or two points which the reader may adjudge worthy of further thought. I am not about to expound or comment on the works of Carl Jung, Sigmund Freud, and others – it is simply not within my competence. However, I thought it worthwhile to share some thoughts that occupy my mind on the subject, and may well resonate with the reader.

Knowledge of Self

Consciousness can be considered a *knowledge of self* as a distinct and separate individual. It is generally accepted in science that nothing can explain itself, yet here we have human consciousness aware of self. Only certain types of living organisms have this awareness: non-living material elements lack this faculty. So the question arises: if individual atoms and molecules are not self-aware, wherein the properties that allow an arrangement or aggregation of molecules to become self-aware? This is not a new question, and I doubt whether it will ever be answered, but it remains a question nonetheless. As previously discussed, *emergent property* is nothing but a label for our ignorance. In all verifiable instances of emergence, the parent entity already contained the seed properties of the properties that later emerged in the child – nothing entirely new arose. Given that there is nothing in the material world that has even a hint of self-awareness, the suggestion that self-awareness arose from the material is implausible.

The other difficulty for emergence is the issue of information and knowledge. To be self-aware means to have knowledge of self, but to have knowledge requires the acquisition of information. Information is derived by processing data through a framework of concepts which provide context, but the data itself results from decoding physical symbols in a storage medium, in this case the brain. Self itself is a *concept*, not the material arrangement of atoms and molecules that instantiate a particular self, and there is nothing in the intrinsic properties of the material that informs an arrangement that it is a unique arrangement: again, nothing can explain itself. To claim that self-awareness emerged from the material, one has to provide a plausible explanation for the origins of data and the mechanisms that provide the requisite functionality.

As has been the focus of this book, no such explanations exist, and in the opinion of this author, they cannot exist, for the material can never belong in the same category or environment as the conceptual.

Control of Self

I remember reading the accounts of prisoners-of-war in various conflicts, and how they coped with the pressures: some endured, others did not. I recall one account of a US serviceman imprisoned in North Vietnam in what came to be called the Hanoi Hilton. He attributed his survival, and the survival of many of his co-prisoners, to the ability to act in a compliant, even subservient, manner toward their guards and interrogators while retaining an inner courage and steadfastness. I ask myself: how can that be if we have no control over self, if we have no volition or free will? The evidence, particularly in these extreme circumstances, is that there exists in the human, a mechanism that allows individual control of response to environment and circumstance.

Loss of Self

By the time Darwin reflected on his life whilst writing his autobiography, the sexagenarian noted an unmistakable sense of loss:

> "I have said that in one respect my mind has changed during the last twenty or thirty years. Up to the age of thirty, or beyond it, poetry of many kinds, such as the works of Milton, Gray, Byron, Wordsworth, Coleridge, and Shelley, gave me great pleasure, and even as a schoolboy I took intense delight in Shakespeare, especially in the historical plays. I have also said that formerly pictures gave me considerable, and music very great delight. But now for many years I cannot endure to read a line of poetry; I have tried lately to read Shakespeare, and found it so intolerably dull that it nauseated me. I have almost lost my taste for pictures and music. Music generally sets me thinking too energetically on what I have been working on, instead of giving me pleasure. I retain some taste for fine scenery, but it does not cause me the exquisite delight which it formerly did."

Darwin concluded by saying:

> "My mind seems to have become a kind of machine for grinding general laws out of large collections of fact, but why this should have caused the atrophy of that part of the

brain alone, on which the higher tastes depend, I cannot conceive....The loss of these tastes is a loss of happiness, and may possibly be injurious to the intellect, and more probably to the moral character, by enfeebling the emotional part of our nature."[97]

"What a sad commentary! But it was not science that did this to Darwin: it was *scientism*. After all, a worldview that sees everything in terms of blind, completely random actions, holding that the only explanations that count are natural processes functioning via unbroken natural laws in non-purposeful ways, would tend to have a negative effect on precisely those areas described by Darwin. While this might not have been an ineluctable outcome, it certainly seems a very probable one. Darwin's physicians said that he died of "heart-failure" and indeed he did, but let me offer that the heart that failed him did so before his death on April 19, 1882." [Author's note: This paragraph is not mine, but I have misplaced the source.]

I believe it reasonable to conclude that Darwin came to understand that there is more to our human existence than evolution can explain. By pursuing evolution with great zeal, Charles Darwin diminished himself, losing his love of life and all that it offers. It is apparent then, to me at least, that evolution not only cannot explain the human condition, but also cannot account for it. Evolution strips one or more layers of self from self, and in that regard, is not only an insufficient explanation but a blight on society.

Time for an Aside

Dr. Granville Sewell is Professor of Mathematics at the University of Texas El Paso. He has written three books on numerical analysis, and is the author of a widely used finite element computer program: this attests to his objectivity. The text reproduced below is an excerpt from his book, *In the Beginning: And Other Essays on Intelligent Design*[98]. I trust that you will understand the relevance to the subject of "self".

"One way to appreciate the problem human consciousness poses for Darwinism or any other mechanical theory of evolution is to ask: Is it possible that computers will someday experience consciousness?

If you believe that a mechanical process such as natural selection could have produced consciousness once, it seems you *can't* say it could never happen again, and it might happen faster now, with intelligent designers helping this time. In fact, most Darwinists probably do believe it could and will happen -- not because they have a higher opinion of computers than I do: Everyone knows that in their most impressive displays of "intelligence," computers are just doing *exactly* what they are told to do, nothing more or less.

They believe it will happen because they have a lower opinion of humans: They simply dumb down the definition of consciousness, and say that if a computer can pass a "Turing test," that is, fool a human at the keyboard in the next room into thinking he is chatting with another human, then the computer has to be considered intelligent, and conscious.

With the right software, my laptop may already be able to pass a Turing test, and convince me that I am Instant Messaging another human. If I type in "My cat died last week" and the computer responds "I am saddened by the death of your cat," and if I'm pretty gullible, that could convince me that I'm talking to another human. But if I look at the software, I might find something like this:

```
if (verb == 'died')
    fprintf(1,'I am saddened by the death of your %s', noun)
end
```

I'm pretty sure there is more to human consciousness than this, and even if my laptop answers all my questions intelligently, I will still doubt there is "someone" inside my

Intel processor who experiences the same consciousness that I do, and who is really saddened by the death of my cat, though I admit I can't prove that there isn't.

I really don't know how to argue "scientifically" with people who believe computers could be conscious. About all I can say is: What about typewriters? Typewriters also do exactly what they are told to do, and have produced some magnificent works of literature. Do you believe that typewriters can also be conscious?

And if you *don't* believe that intelligent engineers could ever cause machines to attain consciousness, how can you believe that random mutations could accomplish this?"

CHAPTER 14

The Fallacy of Materialism

Fallacy - invalid or otherwise faulty reasoning.

In an earlier chapter, I noted the supposedly scientific claim that the universe could create itself out of nothing. In formal terms, this is an example of a self-referential fallacy. If we accept as fact that the universe was somehow created, the assertion could be restated as: *the created one created itself.* Whether or not it was *out of nothing* is immaterial to this logical fallacy, for that is a separate fallacy - a scientific contradiction. It should be obvious that the verb, *create*, meaning to bring into existence, implies the pre-existence of a creator to do the creating, and therefore nothing can create itself.

I will use that line of thinking to demonstrate that *philosophical materialism* is a fallacy of a similar nature. Quoting from the Skeptic's Dictionary:

> "Philosophical materialism (physicalism) is the metaphysical view that there is only one substance in the universe and that substance is physical or material. Materialists believe that spiritual [non-material] substance does not exist. Paranormal, supernatural, or occult phenomena are either delusions or reducible to natural forces."

The interesting aspect of this is that the material monists regularly contradict themselves on this very issue: they demonstrate non-material behaviour as they make their assertions. In this, they are very similar to those who use their free will to assert that there is no such facility as free will, another example of a self-referential fallacy.

In the definition given above, can we replace the words "metaphysical view" with "material evidence" or similar? Firstly, *metaphysics* is a branch of philosophy, not of the hard sciences, and thus deals with the conceptual, not just the physical. If we attempt to replace these words as suggested, then we have a circular reference: *materialism is the material evidence that everything is material*. It is axiomatic that nothing can explain itself, and thus nothing can be evidence that nothing else exists. Any existence is evidence of itself, but cannot refute the existence of something else except in a narrowly defined, closed system. The existence of a rock, for example, cannot refute the existence of water, except in a closed system. When scientists note the existence of rocks and sands on planet Mars, they do not take that as evidence that water does not exist there.

The very curious thing about the materialists' view of science, is that according to materialism, science itself is a self-refuting fallacy. Let me explain, but before doing so, you might ponder this question raised by William Dembski: *"a self-referential paradox: how can knowing subjects composed only of matter know that they are only composed of matter?"*[99]

In science, as defined by the materialists, every entity is described in terms of physical properties and other physical entities. It has to be this way because otherwise, materialism fails. Every interaction of these physical entities is described by mathematical equations. But what describes mathematics? There is nothing identifiably physical that can define or describe mathematics, other than as a set of symbols organised in an intelligent manner to describe scientific observations in a particular way. So if science is described by *mathematics*, and mathematics is described as *science*, it is clearly a self-referential fallacy. Now let us take a step back: the material is always described in terms of physical properties and other physical entities. What physical properties are used to describe or define philosophy, imagination, volition, heroism, and so forth? I am not here referring to the processes that such exhibit, but to the physical reality that they are supposed to have according to the materialists. Describe "a wish" in terms of its physical reality - space, time, energy and matter (if that is all there is). Provide a mathematical equation for the process of *wishing* as distinct from *thinking* or *imagining*.

Here is another perspective worth considering:

> "Yet there is a still deeper problem with elevating materialism to a fundamental principle of science. Democritus glimpsed it. Locke and his fellow empiricists saw it more clearly, namely, that an empirical science never observes matter as such but rather must infer matter from observations ... What stands behind observation? This question admits of no easy answer, in terms of matter or otherwise. The problem is that observation itself cannot tell us what stands behind observation ... There is no way to stand outside the observational act and verify that reality does in fact match up with observation."[100]

Philosophical Materialism *might* be true, but I have yet to find any evidence that would convince me of its truth.

Continuing in the same vein, let me quote from an interesting article entitled "*You Have to Be Conscious to Deny Consciousness, and Other Conundrums*"[101]:

> "To deny that we are mindful creatures, the materialist also has to deny the existence of any realm of abstract concepts that a mind can access. Yet materialism itself is an abstract concept."

To prove to yourself that *materialism* is an *abstract concept*, attempt to define its existence in physical terms. Again, I am not asking for a dictionary definition, nor should anyone accept *an emergent property of the brain* as even approximating a physical description: define materialism as a physical entity in the way that other physical entities are defined.

If you are unable to do so, then perhaps you might begin to suspect the philosophical materialism is indeed a fallacy.

CHAPTER 15

The Reality of Emergence

Emergence: the appearance of new properties or species in the course of development or evolution.

Anticipating that some readers may be aware of conflicting arguments concerning emergence, I thought it useful to extend the discussion in the general context of this study: *information*.

In earlier chapters, I contended that whilst self-organisation does in fact occur in the material world, it only ever does so based on the intrinsic properties of the material itself: the material cannot be the information. William Dembski addresses this issue rather well in his chapter, *The Medium and the Message*[102]. Wherever information is stored, it is always in a pattern, but a pattern is something that is either consciously constructed and/or observed - the physical reality has no concept of the pattern it contains. The physical is what it is, and whether an individual perceives a pattern is up to the individual, just as I described with tree that "looked like" a rabbit - the tree was not aware of what it looked like. On a material medium, such as paper or a computer disk, an almost infinite number of patterns can be arranged without altering the physical reality of the medium: a paper tape will not become a magnetic tape no matter how many patters are applied.

Now this seems all rather obvious, and perhaps silly to even mention it, but it is the essence of the reality. When a snowflake forms, it is the end result of a number of physical processes which can only do what they do. As varied as they are, snowflakes are dependent for their shape upon a finite number of possibilities based on the intrinsic properties of the source medium:

> "the binding properties of the water molecule dictate what angles are permissible, and then the crystal's chance path through the cloud accounts for the unique result: a six-sided marvel that looks like a work of art. But you won't see snowflakes spell out "John loves Mary" -- that kind of purposeful communication requires more than natural law and chance. It requires information."[103]

In a sense, the snowflake can be perceived as an example of emergence: first you had water vapour, now you have a snowflake; the interaction of temperature and pressure caused water vapour to condense into clouds which rose in the atmosphere to colder levels causing the water to freeze, etc., etc.: one thing emerged from another - one set of properties emerged from an earlier set of properties. However, no new information arose as a consequence: it is still just water in one of the physical states that its properties allow, and the process is reversible. That it is reversible reminds me of an argument used by Richard Dawkins in attempting to prove evolution[104]. In a refutation I wrote:

> "In his discussion on guppies (pp. 133–138), the author portrays their colour change behaviour as evidence for evolution, and while it demonstrates LET [adaptation], it cannot be evidence for GTE [General Theory of Evolution], simply because no permanent genetic change occurs. Under the threat of predation, the brightly coloured guppies changed colour to one that camouflaged them in their environment, but when the threat was removed, the rainbow colour returned.

> Since the process was reversible, the experiment demonstrated that the genome of the guppies already contained the information for camouflage, spots, and bright colours, and that this information was not lost in the generations that manifested one or other of the appearance extremes. That selective pressures cause these changes is interesting, but it does not offer substantive evidence for GTE."[105]

Wherever a process is reversible, it cannot be truly said to be an example of emergence (or evolution) because nothing new has been created: all we have is a different perspective being revealed of the same thing.

A similar phenomenon occurs with the magnetism of a piece of ferrous metal, said to be the spontaneous alignment of the magnetic moment of billions of electrons. You might argue that emergence has occurred, non-magnetic to magnetic, but it was only made possible by the intrinsic properties of the metal - the same could not occur with, for example, copper, lead, or silver. Even so, no new information has emerged, and again, the process is reversible. Even when irreversible, it has never been shown that absent of the intervention of intelligence, anything new has ever emerged that cannot be demonstrated to be the natural consequence of the properties inherent in the source materials. If you consider all of the ideas that have "emerged" from humanity over the centuries, most have some reference to a physical reality, but many do not. Even so, the idea itself is information that does not have a physical reality expressed in material terms. Whilst the material can be the medium of information, it is not, itself, information.

For a scientific discussion, with useful examples, on the absurdity of emergence as a creator of information, I would urge the reader to follow up with this article: *Emergence Is Real, but It Can't Replace Information.*[106]

I trust that the reader will evaluate the arguments and counter arguments in that article for themselves, but in the context of the study I have presented here, there is no evidence of information and knowledge arising spontaneously from …. What exactly? It should be clear that the proponents of emergence presuppose evolution and thus use emergence as the mechanism for evolution, a somewhat circular argument. They also fail to acknowledge the significant differences between basic physical structures like snowflakes, chemicals reacting according to their intrinsic properties, biological activities based on pre-specified information in the cell, and intelligent activities by people. Juxtaposing these events may give the appearance of similarity, but such an appearance is false and misleading.

CHAPTER 16

Conclusions

*"Originality consists in thinking for yourself,
and not thinking unlike other people"*
~ J. FitzJames Stephen ~

My intent in this study has been to understand whether cognitive information, and thus knowledge, could arise organically by the undirected evolutionary processes posited by scientists working in the fields of genetics, biochemistry, evolutionary biology, and the like. My conclusion is that it could not - the gulf between the physical and the conceptual cannot be bridged by any of these physical sciences, if for no other reason than the obvious: the conceptual is not material in nature.

A key point has been that of origins.

The senses represent the interface between every living organism and its external reality. For evolution to be true, any and all information acquired by any organism has to be via one or more of those senses. In the trajectory of evolution, the organism had to come to understand the function of those senses, and the nature of the data received therefrom, but the source of that understanding could only be the very communication channel that it did not understand. The senses cannot explain themselves, nor can they communicate their coding structures and protocols to a brain that is not aware of them, other than as an electro-chemical sensation. In a humorous vein, it is a little like the old "Knock, Knock" joke, except that there is no-one at home to ask "Who's there?" The other major problem in this area is that of verification or validation: what could be the mechanism that ensured that what the brain perceived the senses to be telling it actually matched reality?

In the supposed evolution of anatomy and physiology, we have the problem of "knowing that you can". If arms, legs, wings, fins or whatever evolved, what was the process whereby the organism came to understand the existence of the faculty, its function, and how to make it work properly? If the brain controls the muscles, how did the brain come to know the correct commands to send? If by trial and error, the high error rate would argue for deselection before the faculty could ever contribute to survival success. If one faculty evolved from another, such as a wing from a leg, the problem is compounded by the necessity to "unlearn" the previous function. All told, these difficulties present a plausible argument against undirected evolution by chance mutation and natural selection – deselection seems more the order of the day (or eon).

What that says about the truth of the overarching narrative of evolution I will leave to others to conclude for themselves. Similarly, I offer no alternatives to evolution, for that is not my concern in these pages. Perhaps there are variations to undirected evolution, I cannot say, and I choose not to enter that debate lest it detract from the essence of this study.

Years of experience in the field of information processing have given me an insight to how information originates or is derived. Admittedly computers are synthetic in structure whilst the brain is organic, but they are both material and subject to the same laws of physics and chemistry, the latter being very much dependent upon the former. Nothing in the material world outside the human body provides even a hint of how anything physical can devise a method of storing sense events in an organised manner other than in accordance with its intrinsic properties. Such properties can never express a concept external to itself. Even where sense events could be stored, the material itself cannot "know" of its organisation or meaning. We have no evidence that the organic brain, composed as it is of physical material, can behave in any manner different to the same materials found outside the human body – the same laws of physics and chemistry apply.

It is asserted by the material-monists that the non-material cannot influence the material, but if that be true, I would assert that the converse must also be true: the material cannot influence the non-material. Thus if the many phenomena that we confidently believe to be non-material are truly non-material, then according to the evolutionists' mantra, they cannot be the result of evolution. I believe that I have shown that cognitive information is intrinsically non-material, in which case it could never arise by purely material processes. If that be true, then there must be a non-material agency that allows the material and immaterial to interact.

Unless evolutionists can offer a plausible explanation, a step by step process, of how the brain can store sense events in a manner allowing retrieval in a context external to the material itself, i.e. conceptual, then in no way can the overarching narrative of undirected evolution be claimed as proven. I am entirely confident that there can be no physical bridge between the material and the conceptual, and until someone can demonstrate that the conceptual is also material, then the search for truth must go on, even if it means opening doors and exploring avenues that may reveal inconvenient truths.

I have previously asserted that nothing can explain itself: the following quote associated with *Godel's Incompleteness Theorems* is along the same lines. I could claim that these were my own words, but that would be untrue, for unfortunately I have misplaced the source (again). Thus you can take the truth in them as you will:

> *"You can't prove a system of mathematics from within the system, and you can't derive an information-rich pattern from within the pattern. The information in a book, for instance, cannot be derived from the paper and ink used to print it. It's impossible to bootstrap a book from the bare ingredients."*

References

1. Futuyma, Douglas J., *Science on Trial – The Case for Evolution*, Pantheon Books, New York, 1982, p. 5
2. Futuyma, ibid, p. 36
3. Fodor, Jerry and Piatelli-Palmarini, Massimo, *What Darwin Got Wrong*, Picador, New York, 2011
4. Mazur, Suzan, *The Altenberg 16: An Expose of the Evolution Industry*, North Atlantic Books, Berkeley, CA, 2010
5. Meyer, Stephen C., *Signature in the Cell: DNA and the Evidence for Design*, HarperCollins, New York, 2009
6. Sanford, Dr John C., *Genetic Entropy & The Mystery of the Genome*, FMS Publications, Waterloo, New York, 2008
7. Shapiro, James A., *Evolution: A View from the 21st Century*, FT Press Science, Upper Saddle River, NJ, 2013
8. Spetner, Dr. Lee, *The Evolution Revolution - Why People are Rethinking the Theory of Evolution*, Judaica Press, Brooklyn, NY, 2014
9. Wesson, Robert, *Beyond Natural Selection*, A Bradford Book, The MIT Press, Cambridge Massachusetts, 1991
10. Biological Information - New Perspectives, Proceedings of the Symposium *Cornell University, USA, 31 May – 3 June 2011*, Edited by: Robert J Marks II (*Baylor University, USA*), Michael J Behe (*Lehigh University, USA*), William A Dembski (*Discovery Institute, USA*), Bruce L Gordon (*Houston Baptist University, USA*), John C Sanford (*Cornell*)
11. http://www.biologicalinformationnewperspectives.org/#!synopsis/c1294
12. Dembski, William, *Being As Communion - A Metaphysics of Information*, Ashgate Publishing Company, Burlington, VT, 2014

13. Dennett, Daniel C., *Consciousness Explained*, Penguin Press, London, England, 1991
14. Eccles, John C., *How the SELF Controls Its BRAIN*, Springer-Verlag, Berlin, Germany, 1994
15. Gitt, Dr. Werner, *In the Beginning was Information*, First Master Books, Green Forest, AR, 2007
16. Gitt, Dr. Werner, *Without Excuse*, Creation Book Publishers, Atlanta, GA, 2011
17. Heidegger, Martin, *An Introduction to Metaphysics*, Anchor Books, New York, 1961
18. Nagel, Thomas, *Mind and Cosmos: Why the Materialist Neo-Darwinian Conception of Nature is Almost Certainly False*, Oxford University Press, New York, NY, 2012
19. Noble, Denis, *The Music of Life: Biology Beyond Genes*, Oxford University Press, Oxford, UK, 2006
20. Popper, Karl R., and Eccles, John C., *The Self and Its Brain: An Argument for Interactionism*, Routledge & Kegan Paul, London, England, 1983
21. ReMine, Walter James, *The Biotic Message: Evolution Versus Message Theory*, St. Paul Science, Inc., St. Paul, MN, 1993
22. Ryle, Gilbert, *The Concept of Mind*, Penguin Books Ltd, London, England, 1990
23. Collins, Francis S., *The Language of God*, Free Press, Simon & Schuster, New York, 2006
24. Alexander, Denis, *Creation or Evolution: Do We Have To Choose?* Monarch Books, Oxford, UK, 2008
25. Talbot, Wayne, *The Dawkins Deficiency – Why Evolution is Not the Greatest Show on Earth*, Deep River Books, Sisters, OR, 2011
26. Dawkins, Richard, *The Greatest Show on Earth – the Evidence for Evolution*, Bantam Press, London, 2009
27. Kerkut, G.A., *Implications of Evolution*, Pergamon, Oxford, UK, 1960, p 157
28. Wesson, ibid, p. 4
29. Lewontin, Richard, *Billions and billions of demons*, The New York Review, January 9, 1997, p. 31
30. Futuyma, ibid, p. 164
31. Futuyma, ibid, p. 164
32. Taylor, Dr. David, *Business Engineering with Object Technology*, John Wiley & Sons, 1995

33. Hubert, Richard, *Convergent Architecture: Building Model Driven J2EE Systems with UML*, John Wiley & Sons, 2001
34. Ryle, Gilbert, *The Concept of Mind*, Penguin Books Ltd, London, England, 1990
35. O'Leary, Denyse, *Evolution: The Fossils Speak, but Hardly with One Voice*, July 8, 2015, http://www.evolutionnews.org/2015/07/evolution_the_f097491.html
36. Dennett, Daniel C., *Consciousness Explained*, Penguin Press, London, England, 1991
37. Dennett, ibid, pp. 39-40
38. Humphreys, Dr Russell, *New view of gravity explains cosmic microwave background radiation*, Journal of Creation, Vol. 28(3) 2014, pp. 106-114
39. Gitt, Dr. Werner, *In the Beginning was Information*, First Master Books, Green Forest, AR, 2007, p. 60
40. ReMine, Walter James, *The Biotic Message: Evolution Versus Message Theory*, St. Paul Science, Inc., St. Paul, MN, 1993
41. Dembski, William A., *Conservation of Information Made Simple*, 28 Aug 2012, http://www.evolutionnews.org/2012/08/conservation_of063671.html
42. Eccles, John C., *How the SELF Controls Its BRAIN*, Springer-Verlag, Berlin, Germany, 1994, p. 146
43. Eccles, ibid, p. 56
44. Barnstein, Henry, *The Targum of Onkelos to Genesis - A Critical Enquiry*, Isha Books, New Delhi, India, 1896
45. Dawkins, Richard, *The blind watchmaker*, Norton Press, New York NY, 1996, pp. 159-160
46. Andrews, Professor E.H., *Who Made God?*, EP Books, Darlington, England, 2009, p. 157-159
47. http://news.nationalgeographic.com/news/2007/04/photogalleries/giant-crystals-cave/
48. Gitt, ibid
49. Popper, Karl R., and Eccles, John C., *The Self and Its Brain: An Argument for Interactionism*, Routledge & Kegan Paul, London, England, 1983, p. 22
50. Don Miner, Marc Pickett, Marie desJardins, *Understanding the Brain's Emergent Properties*, Department of Computer Science and Electrical Engineering, University of Maryland, Baltimore County, undated
51. CMI-Daily Email AU webmasteraust@creation.info, 30th November, 2015

52 Ross, J., *Bees no drones when it comes to landing*, theaustralian.com, 29 October 2013
53 Baird, E., Boeddeker, N., Ibbotson, M., and Srinivasan, M., *A universal strategy for visually guided landing*, Proceedings of the National Academy of Sciences (USA) **110**(46):18686–18691, 2013
54 Esch, H., Zhang, S., Srinivasan, M.V. and Tautz, J., Honeybee dances communicate distances measured by optic flow, Nature **411**(6837):581–583, 31 May 2001
55 Sarfati, J., *Can it bee? Creation* **25**(2):44–45, 2003; creation.com/bee
56 Ross, ibid
57 Keating, Joseph C. Jnr., *The Meaning of Innate*, Journal of the Canadian Chiropractic Association, **46(1)**, 4-10
58 Eccles, ibid, p. x
59 Eccles, ibid
60 Dawkins, Richard, *The Greatest Show on Earth – the Evidence for Evolution*, Bantam Press, London, 2009, p. 165
61 Talbot, ibid, pp. 100-101
62 Bell, P., *Mudskippers - marvels of the mudflats*, Creation **34**(2), 2012, pp. 48-50
63 Talbot, ibid, pp. 95-96
64 Cashmore, A., *The Lucretian* swerve: The biological basis of human behaviour and the criminal justice system, *Proceedings of the National Academy of Sciences* **107**(10), 2010, p. 4499-4504
65 Cashmore, ibid, p. 4502
66 Johnson, P., *Darwin on Trial*, 2nd ed., Illinois, USA, 1993, p. 127
67 Ryle, ibid, p. 9
68 Naylor, Carma, *A Mormon's Unexpected Journey - Volumes I & 2*, WinePress Publishing, Enumclaw, WA, 2006
69 Coffey, Tony, *Once a Catholic: What You Need to Know about Roman Catholicism*, Harvest House Publishers, Eugene, OR, 1993
70 Rassamni, Jerry, *From Jihad to Jesus – An Ex-Militant's Journey of Faith*, Living Ink Books, Chattanooga, Tennessee, 2006
71 Shoebat, Walid, *Why I Left Jihad*, Top Executive Media, USA, 2005
72 Collins, ibid
73 Flew, Anthony, *There is a God – how the world's most notorious atheist changed his mind*, HarperOne, New York, 2007
74 Hitchens, Peter, *The Rage Against God – How Atheism Led Me to Faith*, Zondervan, Grand Rapids, Michigan, 2010

75 Lewis, C.S., *Surprised by Joy*, Harper Collins, London, 1955
76 Lewis, C.S., *The Case for Christianity*, Collier Books, New York, NY, 1989
77 Sanford, Dr John C., *Genetic Entropy & The Mystery of the Genome*, FMS Publications, Waterloo, New York, 2008
78 Strobel, Lee, *The Case for a Creator*, Zondervan Press, Grand Rapids, MI, 2004
79 Darwish, Nonie, *Now They Call Me Infidel – Why I Renounced Jihad*, Penguin Group (USA), 2007
80 Moreland, J.P., *Scaling the Secular City*, Baker Books, Grand Rapids, MI, 1987, p.83
81 Popper, Karl R., and Eccles, John C., *The Self and Its Brain: An Argument for Interactionism*, Routledge & Kegan Paul, London, England, 1983
82 Popper, ibid, pp. 16-17
83 Popper, ibid
84 Popper, ibid, p. 15
85 Ryle, ibid, p.10
86 Ryle, ibid, p. 13
87 Ryle, ibid, p. 17
88 Dennett, Daniel C., *Consciousness Explained*, Penguin Press, London, England, 1991
89 Dennett, ibid, p. 33
90 Dennett, ibid, p. 35
91 Andrews, Professor E.H., *Who Made God?* EP Books, Carlisle, PA, 2009, p. 57
92 Lewis, C.S., *The Case for Christianity*, Collier Books, New York, NY, 1989, p. 32
93 Burmeister, L.E., *Reading Strategies for Middle and Secondary School Teachers*, Second Edition, Addison-Wesley Pub. Co., Reading, Mass., p.23, 1974
94 Kroeber, A.L., *Anthropology: Culture Patterns and Processes*, Harcourt, Brace, and World, New York, p.7, 1948, p.32
95 Chard, C.S., *Man in Prehistory*, Second Edition, McGraw-Hill Book Co., New York, 1969, p.260
96 Albert Einstein, quoted in Dechant, E., *Reading Improvement in the Secondary School*, Prentice-Hall, Inc., Englewood Cliffs, N.J., p.5, 1973
97 *Life and Letters of Charles Darwin* (1891)) [Public domain]
98 Sewell, Dr. Granville, *In the Beginning: And Other Essays on Intelligent Design*, Discovery Institute Press, Seattle, WA, 2015

[99] Dembski, William, *Being As Communion - A Metaphysics of Information*, Ashgate Publishing Company, Burlington, VT, 2014, p. 7
[100] *Ibid*, p. 81
[101] http://www.evolutionnews.org/2015/07/you_have_to_be097421.html
[102] Dembski, *Ibid*, Chapter 11
[103] http://www.evolutionnews.org/2014/10/emergence_is_re090261.html
[104] Dawkins, Richard, *The Greatest Show on Earth – the Evidence for Evolution*, Bantam Press, London, 2009
[105] Talbot, Wayne, *The Dawkins Deficiency – Why Evolution is Not the Greatest Show on Earth*, Deep River Books, Sisters, OR, 2011, p. 187
[106] http://www.evolutionnews.org/2014/10/emergence_is_re090261.html

About the Author

Wayne Talbot has had a lifelong passion for a travel and a modest appetite for adventure. In addition to exploring his native Australia, he has trekked in the Khumbu and Annapurna regions of Nepal, the Peruvian Andes, Patagonia, and has experienced the beauty of Antarctica. He has motorcycled throughout Australia, and through parts of the USA, New Zealand, Iceland, Bolivia, Peru, Turkey, and North India, with plans for Southeast Europe in 2016. Though a late starter in the literary field, he has written eleven books on matters theological, published on Amazon Kindle under the series, From the Back Pew. Further details can be found on his website, www.peshatbooks.com.

Amongst a variety of interests, he has also pursued the truth of evolution objectively and without seeking refuge in any arguments from theological sources. Either the science holds up, or it does not: what the alternatives may be are irrelevant to the analysis. His first book on this subject, The Dawkins Deficiency, was a refutation of Richard Dawkins's The Greatest Show on Earth. In this work, he sought to demonstrate that much of Dawkins's case was based on a materialistic philosophy rather than scientific fact, commenting, "If evolution theory has been proven, then we should reasonably ask, what specifically has been proven, and what is the evidence?"

His latest study, Information, Knowledge, Evolution and Self, utilises his extensive experience in the information industry, primarily as an analyst in a number of roles. He approaches the subject of evolution from the perspective of the human cognitive ability to acquire information and knowledge, these being conceptual as opposite to the physical nature of information in the genome.

www.ingramcontent.com/pod-product-compliance
Lightning Source LLC
Chambersburg PA
CBHW030752180526
45163CB00003B/998